D0368843

LINEAR INTEGRAL EQUATIONS
Theory and Technique

Linear
Integral Equations
Theory and Technique

Pennsylvania State University

University Park, Pennsylvania

ACADEMIC PRESS   **1971**   New York and London

ACADEMIC PRESS, INC.
111 Fifth Avenue, New York, New York 10003

United Kingdom Edition published by
ACADEMIC PRESS, INC. (LONDON) LTD.
Berkeley Square House, London W1X 6BA

LIBRARY OF CONGRESS CATALOG CARD NUMBER: 77-156268

AMS (MOS) 1970 Subject Classification: 45A05, 45B05, 45C05,
45D05, 45E05, 45E10, 45E99, 45F05, 35C15, 44A25, 44A35, 47G05

PRINTED IN THE UNITED STATES OF AMERICA

IN LOVING MEMORY OF
MY GRANDMOTHER

CONTENTS

PREFACE

Many physical problems which are usually solved by differential equation methods can be solved more effectively by integral equation methods. Indeed, the latter have been appearing in current literature with increasing frequency and have provided solutions to problems heretofore not solvable by standard methods of differential equations. Such problems abound in many applied fields, and the types of solutions explored here will be useful particularly in applied mathematics, theoretical mechanics, and mathematical physics.

Each section of the book contains a selection of examples based on the technique of that section, thus making it possible to use the book as the text for a beginning graduate course. The latter part of the book will be equally useful to research workers who encounter boundary value problems. The level of mathematical knowledge required of the reader is no more than that taught in undergraduate applied mathematics courses. Although no attempt has been made to attain a high level of mathematical rigor, the regularity conditions for every theorem have been stated precisely. To keep the book to a manageable size, a few long proofs have been omitted. They are mostly those proofs which do not appear to be essential in the study of the subject for purposes of applications.

We have omitted topics such as Wiener–Hopf technique and dual integral equations because most of the problems which can be solved by these methods can be solved more easily by the integral equation methods presented in Chapters 10 and 11. Furthermore, we have concentrated mainly on three-dimensional problems. Once the methods have been grasped, the student can readily solve the corresponding plane problems. Besides, the powerful tools of complex variables are available for the latter problems.

The book has developed from courses of lectures on the subject given by the author over a period of years at Pennsylvania State University.

Since it is intended mainly as a textbook we have omitted references to the current research literature. The bibliography contains references to books which the author has consulted and to which the reader is referred occasionally in the text. Chapter 10 is based on author's joint article with Dr. D. L. Jain (*SIAM Journal of Applied Mathematics*, **20**, 1971) while Chapter 11 is based on an article by the author (*Journal of Mathematics and Mechanics*, **19**, 1970, 625–656). The permission of the publishers of these two journals is thankfully acknowledged. These articles contain references to the current research literature pertaining to these chapters. The most important of these references are the research works of Professor W. E. Williams and Professor B. Noble.

Finally, I would like to express my gratitude to my students Mr. P. Gressis, Mr. A. S. Ibrahim, Dr. D. L. Jain, and Mr. B. K. Sachdeva for their assistance in the preparation of the textual material, to Mrs. Suzie Mostoller for her pertinent typing of the manuscript, and to the staff of Academic Press for their helpful cooperation.

LINEAR INTEGRAL EQUATIONS
Theory and Technique

1.1. DEFINITION

An integral equation is an equation in which an unknown function appears under one or more integral signs. Naturally, in such an equation there can occur other terms as well. For example, for $a \leqslant s \leqslant b$, $a \leqslant t \leqslant b$, the equations

$$f(s) = \int_a^b K(s,t)\,g(t)\,dt \ , \tag{1}$$

$$g(s) = f(s) + \int_a^b K(s,t)\,g(t)\,dt \ , \tag{2}$$

$$g(s) = \int_a^b K(s,t)\,[g(t)]^2\,dt \ , \tag{3}$$

where the function $g(s)$ is the unknown function while all the other functions are known, are integral equations. These functions may be complex-valued functions of the real variables s and t.

Integral equations occur naturally in many fields of mechanics and mathematical physics. They also arise as representation formulas for the solutions of differential equations. Indeed, a differential equation can be replaced by an integral equation which incorporates its boundary conditions. As such, each solution of the integral equation automatically satisfies these boundary conditions. Integral equations also form one of the most useful tools in many branches of pure analysis, such as the theories of functional analysis and stochastic processes.

One can also consider integral equations in which the unknown function is dependent not only on one variable but on several variables. Such, for example, is the equation

$$g(s) = f(s) + \int_{\Omega} K(s, t) g(t) \, dt \,, \tag{4}$$

where s and t are n-dimensional vectors and Ω is a region of an n-dimensional space. Similarly, one can also consider systems of integral equations with several unknown functions.

An integral equation is called linear if only linear operations are performed in it upon the unknown function. The equations (1) and (2) are linear, while (3) is nonlinear. In fact, the equations (1) and (2) can be written as

$$L[g(s)] = f(s) \,, \tag{5}$$

where L is the appropriate integral operator. Then, for any constants c_1 and c_2, we have

$$L[c_1 g_1(s) + c_2 g_2(s)] = c_1 L[g_1(s)] + c_2 L[g_2(s)] \,. \tag{6}$$

This is the general criterion for a linear operator. In this book, we shall deal only with linear integral equations.

The most general type of linear integral equation is of the form

$$h(s) g(s) = f(s) + \lambda \int_{a} K(s, t) g(t) \, dt \,, \tag{7}$$

where the upper limit may be either variable or fixed. The functions f, h, and K are known functions, while g is to be determined; λ is a nonzero, real or complex, parameter. The function $K(s, t)$ is called the kernel. The following special cases of equation (7) are of main interest.

(i) FREDHOLM INTEGRAL EQUATIONS.   In all Fredholm integral equations, the upper limit of integration b, say, is fixed.

(i)   In the Fredholm integral equation of the first kind, $h(s) = 0$.

Thus,

$$f(s) + \lambda \int_a^b K(s,t) g(t) \, dt = 0 \,. \tag{8}$$

(ii)   In the Fredholm integral equation of the second kind, $h(s) = 1$;

$$g(s) = f(s) + \lambda \int_a^b K(s,t) g(t) \, dt \,. \tag{9}$$

(iii)   The homogeneous Fredholm integral equation of the second kind is a special case of (ii) above. In this case, $f(s) = 0$;

$$g(s) = \lambda \int_a^b K(s,t) g(t) \, dt \,. \tag{10}$$

(ii)   **VOLTERRA EQUATIONS.**   Volterra equations of the first, homogeneous, and second kinds are defined precisely as above except that $b = s$ is the variable upper limit of integration.

Equation (7) itself is called an integral equation of the third kind.

(iii)   **SINGULAR INTEGRAL EQUATIONS.** When one or both limits of integration become infinite or when the kernel becomes infinite at one or more points within the range of integration, the integral equation is called singular. For example, the integral equations

$$g(s) = f(s) + \lambda \int_{-\infty}^{\infty} (\exp - |s-t|) g(t) \, dt \tag{11}$$

and

$$f(s) = \int_0^s [1/(s-t)^\alpha] g(t) \, dt \,, \qquad 0 < \alpha < 1 \tag{12}$$

are singular integral equations.

1.2. REGULARITY CONDITIONS

We shall be mainly concerned with functions which are either continuous, or integrable or square-integrable. In the subsequent analysis, it will be pointed out what regularity conditions are expected of the functions involved. The notion of Lebesgue-integration is essential to modern mathematical analysis. When an integral sign is used, the Lebesgue integral is understood. Fortunately, if a function is Riemann-integrable, it is also Lebesgue-integrable. There are functions that are Lebesgue-integrable but not Riemann-integrable, but we shall not encounter them in this book. Incidentally, by a square-integrable function $g(t)$, we mean that

$$\int_a^b |g(t)|^2 \, dt < \infty . \tag{1}$$

This is called an $\mathscr{L}_2$-function. The regularity conditions on the kernel $K(s, t)$ as a function of two variables are similar. It is an $\mathscr{L}_2$-function if:

(a) for each set of values of s, t in the square $a \leqslant s \leqslant b \; a \leqslant t \leqslant b$,

$$\int_a^b \int_a^b |K(s, t)|^2 \, ds \, dt < \infty , \tag{2}$$

(b) for each value of s in $a \leqslant s \leqslant b$,

$$\int_a^b |K(s, t)|^2 \, dt < \infty , \tag{3}$$

(c) for each value of t in $a \leqslant t \leqslant b$,

$$\int_a^b |K(s, t)|^2 \, ds < \infty . \tag{4}$$

1.3. SPECIAL KINDS OF KERNELS

(i) SEPARABLE OR DEGENERATE KERNEL. A kernel $K(s, t)$ is called separable or degenerate if it can be expressed as the sum of a finite number of terms, each of which is the product of a function of s only and a function of t only, i.e.,

$$K(s, t) = \sum_{i=1}^n a_i(s) b_i(t) . \tag{1}$$

The functions $a_i(s)$ can be assumed to be linearly independent, otherwise the number of terms in relation (1) can be reduced (by linear independence it is meant that, if $c_1 a_1 + c_2 a_2 + \cdots + c_n a_n = 0$, where c_i are arbitrary constants, then $c_1 = c_2 = \cdots = c_n = 0$).

(ii) SYMMETRIC KERNEL. A complex-valued function $K(s, t)$ is called symmetric (or Hermitian) if $K(s, t) = K^*(t, s)$, where the asterisk denotes the complex conjugate. For a real kernel, this coincides with definition $K(s, t) = K(t, s)$.

1.4. EIGENVALUES AND EIGENFUNCTIONS

If we write the homogeneous Fredholm equation as

$$\int_a^b K(s, t) g(t)\, dt = \mu g(s) , \qquad \mu = 1/\lambda ,$$

we have the classical eigenvalue or characteristic value problem; μ is the eigenvalue and $g(t)$ is the corresponding eigenfunction or characteristic function. Since the linear integral equations are studied in the form (1.1.10), it is λ and not $1/\lambda$ which is called the eigenvalue.

1.5. CONVOLUTION INTEGRAL

Many interesting problems of mechanics and physics lead to an integral equation in which the kernel $K(s, t)$ is a function of the difference $(s - t)$ only:

$$K(s, t) = k(s - t) , \tag{1}$$

where k is a certain function of one variable. The integral equation

$$g(s) = f(s) + \lambda \int_a^s k(s - t) g(t)\, dt , \tag{2}$$

and the corresponding Fredholm equation are called integral equations of the convolution type.

The function defined by the integral

$$\int_0^s k(s-t)g(t)\,dt = \int_0^s k(t)g(s-t)\,dt \tag{3}$$

is called the convolution or the Faltung of the two functions k and g. The integrals occurring in (3) are called the convolution integrals.

The convolution defined by relation (3) is a special case of the standard convolution

$$\int_{-\infty}^\infty k(s-t)g(t)\,dt = \int_{-\infty}^\infty k(t)g(s-t)\,dt . \tag{4}$$

The integrals in (3) are obtained from those in (4) by taking $k(t) = g(t) = 0$, for $t < 0$ and $t > s$.

1.6. THE INNER OR SCALAR PRODUCT OF TWO FUNCTIONS

The inner or scalar product (ϕ,ψ) of two complex $\mathscr{L}_2$-functions ϕ and ψ of a real variable s, $a \leqslant s \leqslant b$, is defined as

$$(\phi,\psi) = \int_a^b \phi(t)\psi^*(t)\,dt . \tag{1}$$

Two functions are called orthogonal if their inner product is zero, that is, ϕ and ψ are orthogonal if $(\phi,\psi) = 0$. The norm of a function $\phi(t)$ is given by the relation

$$\|\phi\| = \left[\int_a^b \phi(t)\phi^*(t)\,dt\right]^{1/2} = \left[\int_a^b |\phi(t)|^2\,dt\right]^{1/2} . \tag{2}$$

A function ϕ is called normalized if $\|\phi\| = 1$. It is obvious that a nonnull function (whose norm is not zero) can always be normalized by dividing it by its norm.

We shall have a great deal more to say about these ideas in Chapter 7. For the time being, we shall content ourselves with mentioning the Schwarz and Minkowski inequalities,

$$|(\phi,\psi)| \leqslant \|\phi\| \, \|\psi\| \qquad (3)$$

and

$$\|\phi+\psi\| \leqslant \|\phi\| + \|\psi\| , \qquad (4)$$

respectively.

1.7. NOTATION

For Fredholm integral equations, it will be assumed that the range of integrations is a to b, unless the contrary is explicitly stated. The quantities a and b will be omitted from the integral sign in the sequel.

2.1. REDUCTION TO A SYSTEM OF ALGEBRAIC EQUATIONS

In Chapter 1, we have defined a degenerate or a separable kernel $K(s, t)$ as

$$K(s, t) = \sum_{i=1}^{n} a_i(s) b_i(t) , \tag{1}$$

where the functions $a_1(s), ..., a_n(s)$ and the functions $b_1(t), ..., b_n(t)$ are linearly independent. With such a kernel, the Fredholm integral equation of the second kind,

$$g(s) = f(s) + \lambda \int K(s, t) g(t) \, dt \tag{2}$$

becomes

$$g(s) = f(s) + \lambda \sum_{i=1}^{n} a_i(s) \int b_i(t) g(t) \, dt . \tag{3}$$

It emerges that the technique of solving this equation is essentially dependent on the choice of the complex parameter λ and on the definition of

$$c_i = \int b_i(t) g(t) \, dt . \tag{4}$$

The quantities c_i are constants, although hitherto unknown.

Substituting (4) in (3) gives

$$g(s) = f(s) + \lambda \sum_{i=1}^{n} c_i a_i(s) , \tag{5}$$

and the problem reduces to finding the quantities c_i. To this end, we put the value of $g(s)$ as given by (5) in (3) and get

$$\sum_{i=1}^{n} a_i(s) \{c_i - \int b_i(t) [f(t) + \lambda \sum_{k=1}^{n} c_k a_k(t)] \, dt\} = 0 . \tag{6}$$

But the functions $a_i(s)$ are linearly independent; therefore,

$$c_i - \int b_i(t) [f(t) + \lambda \sum_{k=1}^{n} c_k a_k(t)] \, dt = 0 , \qquad i = 1, ..., n . \tag{7}$$

Using the simplified notation

$$\int b_i(t) f(t) = f_i , \qquad \int b_i(t) a_k(t) \, dt = a_{ik} , \tag{8}$$

where f_i and a_{ik} are known constants, equation (7) becomes

$$c_i - \lambda \sum_{k=1}^{n} a_{ik} c_k = f_i , \qquad i = 1, ..., n ; \tag{9}$$

that is, a system of n algebraic equations for the unknowns c_i. The determinant $D(\lambda)$ of this system is

$$D(\lambda) = \begin{vmatrix} 1 - \lambda a_{11} & -\lambda a_{12} & \cdots & -\lambda a_{1n} \\ -\lambda a_{21} & 1 - \lambda a_{22} & \cdots & -\lambda a_{2n} \\ \vdots & & & \\ -\lambda a_{n1} & -\lambda a_{n2} & \cdots & 1 - \lambda a_{nn} \end{vmatrix} , \tag{10}$$

which is a polynomial in λ of degree at most n. Moreover, it is not identically zero, since, when $\lambda = 0$, it reduces to unity.

For all values of λ for which $D(\lambda) \neq 0$, the algebraic system (9), and thereby the integral equation (2), has a unique solution. On the other hand, for all values of λ for which $D(\lambda)$ becomes equal to zero, the algerbaic system (9), and with it the integral equation (2), either is insoluble or has an infinite number of solutions. Setting $\lambda = 1/\mu$ in equation (9), we have the eigenvalue problem of matrix theory. The eigenvalues are given by the polynomial $D(\lambda) = 0$. They are also the eigenvalues of our integral equation.

Note that we have considered only the integral equation of the second kind, where alone this method is applicable.

This method is illustrated with the following examples.

2.2. EXAMPLES

Example 1. Solve the Fredholm integral equation of the second kind

$$g(s) = s + \lambda \int_0^1 (st^2 + s^2 t) g(t) \, dt \, . \tag{1}$$

The kernel $K(s, t) = st^2 + s^2 t$ is separable and we can set

$$c_1 = \int_0^1 t^2 g(t) \, dt \, , \qquad c_2 = \int_0^1 tg(t) \, dt \, .$$

Equation (1) becomes

$$g(s) = s + \lambda c_1 s + \lambda c_2 s^2 \, , \tag{2}$$

which we substitute in (1) to obtain the algebraic equations

$$\begin{aligned} c_1 &= \tfrac{1}{4} + \tfrac{1}{4}\lambda c_1 + \tfrac{1}{5}\lambda c_2 \, , \\ c_2 &= \tfrac{1}{3} + \tfrac{1}{3}\lambda c_1 + \tfrac{1}{4}\lambda c_2 \, . \end{aligned} \tag{3}$$

The solution of these equations is readily obtained as

$$c_1 = (60 + \lambda)/(240 - 120\lambda - \lambda^2) \, , \qquad c_2 = 80/(240 - 120\lambda - \lambda^2) \, . \tag{4}$$

From (2) and (4), we have the solution

$$g(s) = [(240 - 60\lambda)s + 80\lambda s^2]/(240 - 120\lambda - \lambda^2) \, . \tag{5}$$

Example 2. Solve the integral equation

$$g(s) = f(s) + \lambda \int_0^1 (s + t) g(t) \, dt \tag{6}$$

and find the eigenvalues.

Here, $a_1(s) = s$, $a_2(s) = 1$, $b_1(t) = 1$, $b_2(t) = t$,

$$a_{11} = \int_0^1 t\,dt = \tfrac{1}{2}, \qquad a_{12} = \int_0^1 dt = 1,$$

$$a_{21} = \int_0^1 t^2\,dt = \tfrac{1}{3}, \qquad a_{22} = \int_0^1 t\,dt = \tfrac{1}{2},$$

$$f_1 = \int_0^1 f(t)\,dt, \qquad f_2 = \int_0^1 tf(t)\,dt.$$

Substituting these values in (2.1.9), we have the algebraic system

$$(1-\tfrac{1}{2}\lambda)c_1 - \lambda c_2 = f_1, \qquad -\tfrac{1}{3}\lambda c_1 + (1-\tfrac{1}{2}\lambda)c_2 = f_2.$$

The determinant $D(\lambda) = 0$ gives $\lambda^2 + 12\lambda - 12 = 0$. Thus, the eigenvalues are

$$\lambda_1 = (-6 + 4\sqrt{3}), \qquad \lambda_2 = (-6 - 4\sqrt{3}).$$

For these two values of λ, the homogeneous equation has a nontrivial solution, while the integral equation (6) is, in general, not soluble. When λ differs from these values, the solution of the above algebraic system is

$$c_1 = [-12f_1 + \lambda(6f_1 - 12f_2)]/(\lambda^2 + 12\lambda - 12),$$

$$c_2 = [-12f_2 - \lambda(4f_1 - 6f_2)]/(\lambda^2 + 12\lambda - 12).$$

Using the relation (2.1.5), there results the solution

$$g(s) = f(s) + \lambda \int_0^1 \frac{6(\lambda-2)(s+t) - 12\lambda st - 4\lambda}{\lambda^2 + 12\lambda - 12} f(t)\,dt. \tag{7}$$

The function $\Gamma(s, t; \lambda)$,

$$\Gamma(s, t; \lambda) = [6(\lambda-2)(s+t) - 12\lambda st - 4\lambda]/(\lambda^2 + 12\lambda - 12), \tag{8}$$

is called the resolvent kernel. We have therefore succeeded in inverting the integral equation because the right-hand side of the above formula is a known quantity.

Example 3.   Invert the integral equation

$$g(s) = f(s) + \lambda \int_0^{2\pi} (\sin s \cos t) g(t) \, dt \, . \tag{9}$$

As in the previous examples, we set

$$c = \int_0^{2\pi} (\cos t) g(t) \, dt$$

to obtain

$$g(s) = f(s) + \lambda c \sin s \, . \tag{10}$$

Multiply both sides of this equation by $\cos s$ and integrate from 0 to 2π. This gives

$$c = \int_0^{2\pi} (\cos t) f(t) \, dt \, . \tag{11}$$

From (10) and (11), we have the required formula:

$$g(s) = f(s) + \lambda \int_0^{2\pi} (\sin s \cos t) f(t) \, dt \, . \tag{12}$$

Example 4.   Find the resolvent kernel for the integral equation

$$g(s) = f(s) + \lambda \int_{-1}^{1} (st + s^2 t^2) g(t) \, dt \, . \tag{13}$$

For this equation,

$$a_1(s) = s \, , \qquad a_2(s) = s^2 \, , \qquad b_1(t) = t \, , \qquad b_2(t) = t^2 \, ,$$

$$a_{11} = \tfrac{2}{3} \, , \qquad a_{12} = a_{21} = 0 \, , \qquad a_{22} = \tfrac{2}{5} \, ,$$

$$f_1 = \int_{-1}^{1} t f(t) \, dt \, , \qquad f_2 = \int_{-1}^{1} t^2 f(t) \, dt \, .$$

Therefore, the corresponding algebraic system is

$$c_1(1 - \tfrac{2}{3}\lambda) = f_1 \, , \qquad c_2(1 - \tfrac{2}{5}\lambda) = f_2 \, . \tag{14}$$

Substituting the values of c_1 and c_2 as obtained from (14) in (2.1.9) yields the solution

$$g(s) = f(s) + \lambda \int_{-1}^{1} \left[\frac{st}{(1-\frac{2}{3}\lambda)} + \frac{s^2 t^2}{(1-\frac{2}{5}\lambda)} \right] f(t) \, dt \; . \tag{15}$$

Thus, the resolvent kernel is

$$\Gamma(s,t;\lambda) = \frac{st}{1-(2\lambda/3)} + \frac{s^2 t^2}{1-(2\lambda/5)} \; . \tag{16}$$

We shall now give examples of homogeneous integral equations.

Example 5. Solve the homogeneous Fredholm integral equation

$$g(s) = \lambda \int_{0}^{1} e^s e^t g(t) \, dt \; . \tag{17}$$

Define

$$c = \int_{0}^{1} e^t g(t) \, dt \; ,$$

so that (17) becomes

$$g(s) = \lambda c e^s \; . \tag{18}$$

Put this value of $g(s)$ in (17) and get

$$\lambda c e^s = \lambda e^s \int_{0}^{1} e^t [\lambda c e^t] \, dt = \tfrac{1}{2}\lambda^2 \, e^s c (e^2 - 1)$$

or

$$\lambda c \{2 - \lambda(e^2 - 1)\} = 0 \; .$$

If $c = 0$ or $\lambda = 0$, then we find that $g \equiv 0$. Assume that neither $c = 0$ nor $\lambda = 0$; then, we have the eigenvalue

$$\lambda = 2/(e^2 - 1) \; .$$

Only for this value of λ does there exist a nontrivial solution of the integral equation (17). This solution is found from (18) to be

$$g(s) = [2c/(e^2 - 1)] e^s \; . \tag{19}$$

Thus, to the eigenvalue $2/(e^2 - 1)$ there corresponds the eigenfunction e^s.

Example 6. Find the eigenvalues and eigenfunctions of the homogeneous integral equation

$$g(s) = \lambda \int_1^2 [st + (1/st)] g(t) \, dt .$$

(20)

Comparing this with (2.1.3), we have

$$a_1(s) = s , \qquad a_2(s) = 1/s , \qquad b_1(t) = t , \qquad b_2(t) = 1/t .$$

$$a_{11} = \tfrac{7}{3} , \qquad a_{12} = a_{21} = 1 , \qquad a_{22} = \tfrac{1}{2} .$$

The formula (2.1.9) then yields the following homogeneous equations:

$$(1 - \tfrac{7}{3}\lambda) c_1 - \lambda c_2 = 0 , \qquad -\lambda c_1 + (1 - \tfrac{1}{2}\lambda) c_2 = 0 ,$$

(21)

which have a nontrivial solution only if the determinant

$$D(\lambda) = \begin{vmatrix} 1 - \tfrac{7}{3}\lambda & -\lambda \\ -\lambda & 1 - \lambda \end{vmatrix} = 1 - (17/6)\lambda + (1/6)\lambda^2 ,$$

vanishes. Therefore, the required eigenvalues are

$$\lambda_1 = \tfrac{1}{2}(17 + \sqrt{265}) \simeq 16.6394 ;$$
$$\lambda_2 = \tfrac{1}{2}(17 - \sqrt{265}) \simeq 0.3606 .$$

(22)

The solution corresponding to λ_1 is $c_2 \simeq -2.2732 c_1$, while that corresponding to λ_2 is $c_2' \simeq 0.4399 c_1'$. The eigenfunctions of the given integral equation are found by substituting in (2.1.5):

$$g_1(s) \simeq 16.639 c_1 [s - 2.2732(1/s)] ,$$
$$g_2(s) \simeq 0.3606 c_1' [s + 0.4399(1/s)] ,$$

(23)

where c_1 and c_1' are two undetermined constants.

2.3. FREDHOLM ALTERNATIVE

In the previous sections, we have found that, if the kernel is separable, the problem of solving an integral equation of the second kind reduces to that of solving an algebraic system of equations. Unfortunately, integral equations with degenerate kernels do not occur frequently in practice. But since they are easily treated and, furthermore, the results derived for such equations lead to a better understanding of integral equations of more general types, it is worthwhile to study them. Last, but not least, any reasonably well-behaved kernel can be written as an infinite series of degenerate kernels.

When an integral equation cannot be solved in closed form, then recourse has to be taken to approximate methods. But these approximate methods can be applied with confidence only if the existence of the solution is assured in advance. The Fredholm theorems explained in this chapter provide such an assurance. The basic theorems of the general theory of integral equations, which were first presented by Fredholm, correspond to the basic theorems of linear algebraic systems. Fredholm's classical theory shall be presented in Chapter 4 for general kernels. Here, we shall deal with degenerate kernels and borrow the results of linear algebra.

In Section 2.1, we have found that the solution of the present problem rests on the investigation of the determinant (2.1.10) of the coefficients of the algebraic system (2.1.9). If $D(\lambda) \neq 0$, then that system has only one solution, given by Cramer's rule

$$c_i = (D_{1i}f_1 + D_{2i}f_2 + \cdots + D_{ni}f_n)/D(\lambda) , \qquad i = 1, 2, \cdots, n , \quad (1)$$

where D_{hi} denotes the cofactor of the (h, i)th element of the determinant (2.1.10). Consequently, the integral equation (2.1.2) has the unique solution (2.1.5), which, in view of (1), becomes

$$g(s) = f(s) + \lambda \sum_{i=1}^{n} \frac{D_{1i}f_1 + D_{2i}f_2 + \cdots + D_{ni}f_n}{D(\lambda)} a_i(s) , \qquad (2)$$

while the corresponding homogeneous equation

$$g(s) = \lambda \int K(s, t) g(t) \, dt \qquad (3)$$

has only the trivial solution $g(s) = 0$.

Substituting for f_i from (2.1.8) in (2), we can write the solution $g(s)$ as

$$g(s) = f(s) + [\lambda/D(\lambda)]$$

$$\times \int \left\{ \sum_{i=1}^{n} [D_{1i}b_1(t) + D_{2i}b_2(t) + \cdots + D_{ni}b_n(t)] a_i(s) \right\} f(t) \, dt.$$

$$(4)$$

Now consider the determinant of $(n+1)$th order

$$D(s,t;\lambda) = - \begin{vmatrix} 0 & a_1(s) & a_2(s) & \cdots & a_n(s) \\ b_1(t) & 1-\lambda a_{11} & -\lambda a_{12} & \cdots & -\lambda a_{1n} \\ b_2(t) & -\lambda a_{21} & 1-\lambda a_{22} & \cdots & -\lambda a_{2n} \\ \vdots & & & & \\ b_n(t) & -\lambda a_{n1} & -\lambda a_{n2} & \cdots & 1-\lambda a_{nn} \end{vmatrix}. \tag{5}$$

By developing it by the elements of the first row and the corresponding minors by the elements of the first column, we find that the expression in the brackets in equation (4) is $D(s,t;\lambda)$. With the definition

$$\Gamma(s,t;\lambda) = D(s,t;\lambda)/D(\lambda), \tag{6}$$

equation (4) takes the simple form

$$g(s) = f(s) + \lambda \int \Gamma(s,t;\lambda) f(t) \, dt. \tag{7}$$

The function $\Gamma(s,t;\lambda)$ is the resolvent (or reciprocal) kernel we have already encountered in Examples 2 and 4 in the previous section. We shall see in Chapter 4 that the formula (6) has many important consequences. For the time being, we content ourselves with the observation that the only possible singular points of $\Gamma(s,t;\lambda)$ in the λ-plane are the roots of the equation $D(\lambda) = 0$, i.e., the eigenvalues of the kernel $K(s,t)$.

The above discussion leads to the following basic Fredholm theorem.

Fredholm Theorem.   The inhomogeneous Fredholm integral equation (2.1.2) with a separable kernel has one and only one solution, given by formula (7). The resolvent kernel $\Gamma(s,t;\lambda)$ coincides with the quotient (6) of two polynomials.

If $D(\lambda) = 0$, then the inhomogeneous equation (2.1.2) has no solution in general, because an algebraic system with vanishing determinant can be solved only for some particular values of the quantities f_i. To discuss this case, we write the algebraic system (2.1.9) as

$$(\mathbf{I} - \lambda \mathbf{A})\mathbf{c} = \mathbf{f}, \tag{8}$$

where $\mathbf{I}$ is the unit (or identity) matrix of order n and $\mathbf{A}$ is the matrix (a_{ij}). Now, when $D(\lambda) = 0$, we observe that for each nontrivial solution of the homogeneous algebraic system

$$(\mathbf{I} - \lambda \mathbf{A})\mathbf{c} = 0 \tag{9}$$

there corresponds a nontrivial solution (an eigenfunction) of the homogeneous integral equation (3). Furthermore, if λ coincides with a certain eigenvalue λ_0 for which the determinant $D(\lambda_0) = |\mathbf{I} - \lambda_0 \mathbf{A}|$ has the rank p, $1 \leqslant p \leqslant n$, then there are $r = n - p$ linearly independent solutions of the algebraic system (9); r is called the index of the eigenvalue λ_0. The same holds for the homogeneous integral equation (3). Let us denote these r linearly independent solutions as $g_{01}(s), g_{02}(s), \cdots, g_{0r}(s)$, and let us also assume that they have been normalized. Then, to each eigenvalue λ_0 of index $r = n - p$, there corresponds a solution $g_0(s)$ of the homogeneous integral equation (3) of the form

$$g_0(s) = \sum_{k=1}^{r} \alpha_k g_{0k}(s) \,,$$

where α_k are arbitrary constants.

Let m be the multiplicity of the eigenvalue λ_0, i.e., $D(\lambda) = 0$ has m equal roots λ_0. Then, we infer from the theory of linear algebra that, by using the elementary transformations on the determinant $|\mathbf{I} - \lambda \mathbf{A}|$, we shall have at most $m + 1$ identical rows and this maximum is achieved only if $\mathbf{A}$ is symmetric. This means that the rank p of $D(\lambda_0)$ is greater than or equal to $n - m$. Thus,

$$r = n - p \leqslant n - (n-m) = m \,,$$

and the equality holds only when $a_{ij} = a_{ji}$.

Thus we have proved the theorem of Fredholm that, if $\lambda = \lambda_0$ is a root of multiplicity $m \geqslant 1$ of the equation $D(\lambda) = 0$, then the homogeneous integral equation (3) has r linearly independent solutions; r is the index of the eigenvalue such that $1 \leqslant r \leqslant m$.

The numbers r and m are also called the geometric multiplicity and algebraic multiplicity of λ_0, respectively. From the above result, it follows that the algebraic multiplicity of an eigenvalue must be greater than or equal to its geometric multiplicity.

To study the case when the inhomogeneous Fredholm integral equation (2.1.2) has solutions even when $D(\lambda) = 0$, we need to define and study the transpose of the equation (2.1.2). The integral equation[1]

$$\psi(s) = f(s) + \lambda \int K(t,s) \psi(t) \, dt \tag{10}$$

---

[1] We shall consider only real functions here even though the results are easily extended to complex functions. Outcome is the same in both the cases.

is called the transpose (or adjoint) of the equation (2.1.2). Observe that the relation between (2.1.2) and its transpose (10) is symmetric, since (2.1.2) is the transpose of (10).

If the separable kernel $K(s, t)$ has the expansion (2.1.1), then the kernel $K(t, s)$ of the transposed equation has the expansion

$$K(t, s) = \sum_{i=1}^{n} a_i(t) b_i(s) . \tag{11}$$

Proceeding as in Section 2.1, we end up with the algebraic system

$$(\mathbf{I} - \lambda \mathbf{A}^{\mathrm{T}}) \mathbf{c} = \mathbf{f} , \tag{12}$$

where $\mathbf{A}^{\mathrm{T}}$ stands for the transpose of $\mathbf{A}$ and where c_i and f_i are now defined by the relations

$$c_i = \int a_i(t) g(t) \, dt , \qquad f_i = \int a_i(t) f(t) \, dt . \tag{13}$$

The interesting feature of the system (12) is that the determinant $D(\lambda)$ is the same function as (2.1.10) except that there has been an interchange of rows and columns in view of the interchange in the functions a_i and b_i. Thus, the eigenvalues of the transposed integral equation are the same as those of the original equation. This means that *the transposed equation* (10) *also possesses a unique solution whenever* (2.1.2) *does*.

As regards the eigenfunctions of the homogeneous system

$$|\mathbf{I} - \lambda \mathbf{A}^{\mathrm{T}}| \mathbf{c} = 0 , \tag{14}$$

we know from linear algebra that these are different from the corresponding eigenfunctions of the system (9). The same applies to the eigenfunctions of the transposed integral equation. Since the index r of λ_0 is the same in both these systems, the number of linearly independent eigenfunctions is also r for the transposed system. Let us denote them by $\psi_{01}, \psi_{02}, \cdots, \psi_{0r}$ and let us assume that they have been normalized. Then, any solution $\psi_0(s)$ of the transposed homogeneous integral equation

$$\psi(s) = \lambda \int K(t, s) \psi(t) \, dt \tag{15}$$

corresponding to the eigenvalue λ_0 is of the form

$$\psi_0(s) = \sum \beta_i \psi_{0i}(s) ,$$

where β_i are arbitrary constants.

We prove in passing that eigenfunctions $g(s)$ and $\psi(s)$ corresponding to distinct eigenvalues λ_1 and λ_2, respectively, of the homogeneous integral equation (3) and its transpose (15) are orthogonal. In fact, we have

$$g(s) = \lambda_1 \int K(s,t) g(t) \, dt \, , \qquad \psi(s) = \lambda_2 \int K(t,s) \psi(t) \, dt.$$

Multiplying both sides of the first equation by $\lambda_2 \psi(s)$ and those of the second equation by $\lambda_1 g(s)$, integrating, and then subtracting the resulting equations, we obtain

$$(\lambda_2 - \lambda_1) \int_a^b g(s) \psi(s) \, ds = 0 \, .$$

But $\lambda_1 \neq \lambda_2$, and the result follows.

We are now ready to discuss the solution of the inhomogeneous Fredholm integral equation (2.1.2) for the case $D(\lambda) = 0$. In fact, we can prove that the necessary and sufficient condition for this equation to have a solution for $\lambda = \lambda_0$, a root of $D(\lambda) = 0$, is that $f(s)$ be orthogonal to the r eigenfunctions ψ_{0i} of the transposed equation (15).

The necessary part of the proof follows from the fact that, if equation (2.1.2) for $\lambda = \lambda_0$ admits a certain solution $g(s)$, then

$$\int f(s) \psi_{0i}(s) \, ds = \int g(s) \psi_{0i}(s) \, ds$$

$$- \lambda_0 \int \psi_{0i}(s) \, ds \int K(s,t) g(t) \, dt$$

$$= \int g(s) \psi_{0i}(s) \, ds$$

$$- \lambda_0 \int g(t) \, dt \int K(s,t) \psi_{0i}(s) \, ds = 0 \, ,$$

because λ_0 and $\psi_{0i}(s)$ are eigenvalues and corresponding eigenfunctions of the transposed equation.

To prove the sufficiency of this condition, we again appeal to linear algebra. In fact, the corresponding condition of orthogonality for the linear-algebraic system assures us that the inhomogeneous system (8) reduces to only $n-r$ independent equations. This means that the rank of the matrix $(\mathbf{I} - \lambda \mathbf{A})$ is exactly $p = n-r$, and therefore the system (8) or (2.1.9) is soluble. Substituting this solution in (2.1.5), we have the solution to our integral equation.

Finally, the difference of any two solutions of (2.1.2) is a solution of the homogeneous equation (3). Hence, the most general solution of the inhomogeneous integral equation (2.1.2) has the form

$$g(s) = G(s) + \alpha_1 g_{01}(s) + \alpha_2 g_{02}(s) + \cdots + \alpha_r g_{0r}(s) , \qquad (16)$$

where $G(s)$ is a suitable linear combination of the functions $a_1(s)$, $a_2(s), \cdots, a_n(s)$.

We have thus proved the theorem that, if $\lambda = \lambda_0$ is a root of multiplicity $m \geq 1$ of the equation $D(\lambda) = 0$, then the inhomogeneous equation has a solution if and only if the given function $f(s)$ is orthogonal to all the eigenfunctions of the transposed equation.

The results of this section can be collected to establish the following theorem.

Fredholm Alternative Theorem.
Either the integral equation

$$g(s) = f(s) + \lambda \int K(s,t) g(t) \, dt \qquad (17)$$

with fixed λ possesses one and only one solution $g(s)$ for arbitrary $\mathscr{L}_2$-functions $f(s)$ and $K(s,t)$, in particular the solution $g = 0$ for $f = 0$; or the homogeneous equation

$$g(s) = \lambda \int K(s,t) g(t) \, dt \qquad (18)$$

possesses a finite number r of linearly independent solutions g_{0i}, $i = 1, 2, \cdots, r$. In the first case, the transposed inhomogeneous equation

$$\psi(s) = f(s) + \lambda \int K(t,s) \psi(t) \, dt \qquad (19)$$

also possesses a unique solution. In the second case, the transposed homogeneous equation

$$\psi(s) = \lambda \int K(t,s) \psi(t) \, dt \qquad (20)$$

also has r linearly independent solutions ψ_{0i}, $i = 1, 2, \cdots, r$; the inhomogeneous integral equation (7) has a solution if and only if the given function $f(s)$ satisfies the r conditions

$$(f, \psi_{0i}) = \int f(s) \psi_{0i}(s) \, ds = 0 , \qquad i = 1, 2, \cdots, r . \qquad (21)$$

In this case, the solution of (17) is determined only up to an additive linear combination $\sum_{i=1}^{r} c_i g_{0i}$.

The following examples illustrate the theorems of this section.

2.4. EXAMPLES

Example 1.   Show that the integral equation

$$g(s) = f(s) + (1/\pi) \int_0^{2\pi} [\sin(s+t)] g(t) \, dt \qquad (1)$$

possesses no solution for $f(s) = s$, but that it possesses infinitely many solutions when $f(s) = 1$.

For this equation,

$$K(s, t) = \sin s \cos t + \cos s \sin t \,,$$

$$a_1(s) = \sin s \,, \qquad a_2(s) = \cos s \,, \qquad b_1(t) = \cos t \,, \qquad b_2(t) = \sin t \,.$$

Therefore,

$$a_{11} = \int_0^{2\pi} \sin t \cos t \, dt = 0 = a_{22} \,,$$

$$a_{12} = \int_0^{2\pi} \cos^2 t \, dt = \pi = a_{21} \,.$$

$$D(\lambda) = \begin{vmatrix} 1 & -\lambda\pi \\ -\lambda\pi & 1 \end{vmatrix} = 1 - \lambda^2 \pi^2 \,. \qquad (2)$$

The eigenvalues are $\lambda_1 = 1/\pi$, $\lambda_2 = -1/\pi$ and equation (1) contains $\lambda_1 = 1/\pi$. Therefore, we have to examine the eigenfunctions of the transposed equation (note that the kernel is symmetric)

$$g(s) = (1/\pi) \int_0^{2\pi} \sin(s+t) g(t) \, dt \,. \qquad (3)$$

The algebraic system corresponding to (3) is

$$c_1 - \lambda\pi c_2 = 0 \,, \qquad -\lambda\pi c_1 + c_2 = 0 \,,$$

which gives

$$c_1 = c_2 \quad \text{for} \quad \lambda_1 = 1/\pi \,; \qquad c_1 = -c_2 \quad \text{for} \quad \lambda_2 = -1/\pi \,.$$

Therefore, the eigenfunctions for $\lambda_1 = 1/\pi$ follow from the relation (2.1.5) and are given by

$$g(s) = c(\sin s + \cos s) . \tag{4}$$

Since

$$\int_0^{2\pi} (s \sin s + s \cos s) \, ds = -2\pi \neq 0 ,$$

while

$$\int_0^{2\pi} (\sin s + \cos s) \, ds = 0 ,$$

we have proved the result.

Example 2.   Solve the integral equation

$$g(s) = f(s) + \lambda \int_0^1 (1 - 3st) g(t) \, dt . \tag{5}$$

The algebraic system (2.1.9) for this equation is

$$(1 - \lambda) c_1 + \tfrac{3}{2} \lambda c_2 = f_1 , \qquad -\tfrac{1}{2} \lambda c_1 + (1 + \lambda) c_2 = f_2 , \tag{6}$$

while

$$D(\lambda) = \begin{vmatrix} 1 - \lambda & \tfrac{3}{2}\lambda \\ -\tfrac{1}{2}\lambda & 1 + \lambda \end{vmatrix} = \tfrac{1}{4}(4 - \lambda^2) . \tag{7}$$

Therefore, the inhomogeneous equation (5) will have a unique solution if and only if $\lambda \neq \pm 2$. Then the homogeneous equation

$$g(s) = \lambda \int_0^1 (1 - 3st) g(t) \, dt \tag{8}$$

has only the trivial solution.

Let us now consider the case when λ is equal to one of the eigenvalues and examine the eigenfunctions of the transposed homogeneous equation

$$g(s) = \lambda \int_0^1 (1 - 3st) g(t) \, dt . \tag{9}$$

For $\lambda = +2$, the algebraic system (6) gives $c_1 = 3c_2$. Then, (2.1.5) gives the eigenfunction

$$g(s) = c(1 - s) , \tag{10}$$

where c is an arbitrary constant. Similarly, for $\lambda = -2$, the corresponding eigenfunction is

$$g(s) = c(1-3s) . \tag{11}$$

It follows from the above analysis that the integral equation

$$g(s) = f(s) + 2 \int_0^1 (1-3st) g(t) \, dt$$

will have a solution if $f(s)$ satisfies the condition

$$\int_0^1 (1-s)f(s) \, ds = 0 ,$$

while the integral equation

$$g(s) = f(s) - 2 \int_0^1 (1-3st) g(t) \, dt$$

will have a solution if the following holds:

$$\int_0^1 (1-3s)f(s) \, ds = 0 .$$

2.5. AN APPROXIMATE METHOD

The method of this chapter is useful in finding approximate solutions of certain integral equations. We illustrate it by the following example:

$$g(s) = e^s - s - \int_0^1 s(e^{st} - 1) g(t) \, dt . \tag{1}$$

Let us approximate the kernel by the sum of the first three terms in its Taylor series:

$$K(t, s) = s(e^{st} - 1) \simeq s^2 t + \tfrac{1}{2}s^3 t^2 + \tfrac{1}{6}s^4 t^3 , \tag{2}$$

that is, by a separable kernel. Then the integral equation takes the form

$$g(s) = e^s - s - \int_0^1 (s^2 t + \tfrac{1}{2}s^3 t^2 + \tfrac{1}{6}s^4 t^3) g(t) \, dt . \tag{3}$$

Since the kernel is separable, we require the solution in the form

$$g(s) = e^s - s + c_1 s^2 + c_2 s^3 + c_3 s^4 . \tag{4}$$

Following the method of this chapter, we find that the constants c_1, c_2, c_3 satisfy the following algebraic system:

$$(5/4)c_1 + (1/5)c_2 + (1/6)c_3 = -2/3 ,$$
$$(1/5)c_1 + (13/6)c_2 + (1/7)c_3 = (9/4) - e , \tag{5}$$
$$(1/6)c_1 + (1/7)c_2 + (49/8)c_3 = 2e - (29/5) ,$$

whose solution is

$$c_1 = -0.5010 , \qquad c_2 = -0.1671 , \qquad c_3 = -0.0422 . \tag{6}$$

Substituting these values in (4), we have the solution

$$g(s) = e^s - s - 0.5010s^2 - 0.1671s^3 - 0.0423s^4 . \tag{7}$$

Now the exact solution of the integral equation (1) is

$$g(s) = 1 . \tag{8}$$

Using the approximate solution for $s = 0$, $s = 0.5$, and $s = 1.0$, the value of $g(s)$ from (7) is

$$g(0) = 1.0000 , \qquad g(0.5) = 1.0000 , \qquad g(1) = 1.0080 , \tag{9}$$

which agrees with (8) rather closely.

In Chapter 7 (see Section 7.7), we shall prove that an arbitrary $\mathscr{L}_2$-kernel can be approximated in norm by a separable kernel.

EXERCISES

1. Consider the equation

$$g(s) = f(s) + \lambda \int_0^1 K(s, t) g(t) \, dt$$

and show that for the kernels given in Table I the function $D(\lambda)$ has the given expression.

TABLE I

Case	Kernel	$D(\lambda)$
(i)	± 1	$1 \mp \lambda$
(ii)	st	$1 - (\lambda/3)$
(iii)	$s^2 + t^2$	$1 - (2\lambda/3) - (4\lambda^2/45)$

2. Solve the integral equation

$$g(s) = f(s) + \lambda \int_0^{2\pi} \cos(s+t)\, g(t)\, dt$$

and find the condition that $f(s)$ must satisfy in order that this equation has a solution when λ is an eigenvalue. Obtain the general solution if $f(s) = \sin s$, considering all possible cases.

3. Show that the integral equation

$$g(s) = \lambda \int_0^{\pi} (\sin s \sin 2t)\, g(t)\, dt$$

has no eigenvalues.

4. Solve the integral equation

$$g(s) = 1 + \lambda \int_{-\pi}^{\pi} e^{i\omega(s-t)}\, g(t)\, dt ,$$

considering separately all the exceptional cases.

5. In the integral equation

$$g(s) = s^2 + \int_0^1 (\sin st)\, g(t)\, dt ,$$

replace $\sin st$ by the first two terms of its power-series development

$$\sin st = st - \frac{(st)^3}{3!} + \cdots$$

and obtain an approximate solution.

METHOD OF SUCCESSIVE APPROXIMATIONS

3.1. ITERATIVE SCHEME

Ordinary first-order differential equations can be solved by the well-known Picard method of successive approximations. An iterative scheme based on the same principle is also available for linear integral equations of the second kind:

$$g(s) = f(s) + \lambda \int K(s,t) g(t) \, dt \, . \tag{1}$$

In this chapter, we present this method. We assume the functions $f(s)$ and $K(s,t)$ to be $\mathscr{L}_2$-functions as defined in Chapter 1.

As a zero-order approximation to the desired function $g(s)$, the solution $g_0(s)$,

$$g_0(s) = f(s) \, , \tag{2}$$

is taken. This is substituted into the right side of equation (1) to give the first-order approximation

$$g_1(s) = f(s) + \lambda \int K(s,t) g_0(t) \, dt \, . \tag{3}$$

This function, when substituted into (1), yields the second approximation. This process is then repeated; the $(n+1)$th approximation is obtained by substituting the nth approximation in the right side of (1). There results the recurrence relation

$$g_{n+1}(s) = f(s) + \lambda \int K(s, t) g_n(t) \, dt .\tag{4}$$

If $g_n(s)$ tends uniformly to a limit as $n \to \infty$, then this limit is the required solution. To study such a limit, let us examine the iterative procedure (4) in detail. The first- and second-order approximations are

$$g_1(s) = f(s) + \lambda \int K(s, t) f(t) \, dt\tag{5}$$

and

$$g_2(s) = f(s) + \lambda \int K(s, t) f(t) \, dt$$
$$+ \lambda^2 \int K(s, t) \left[\int K(t, x) f(x) \, dx \right] dt .\tag{6}$$

This formula can be simplified by setting

$$K_2(s, t) = \int K(s, x) K(x, t) \, dx\tag{7}$$

and by changing the order of integration. The result is

$$g_2(s) = f(s) + \lambda \int K(s, t) f(t) \, dt + \lambda^2 \int K_2(s, t) f(t) \, dt .\tag{8}$$

Similarly,

$$g_3(s) = f(s) + \lambda \int K(s, t) f(t) \, dt$$
$$+ \lambda^2 \int K_2(s, t) f(t) \, dt + \lambda^3 \int K_3(s, t) f(t) \, dt ,\tag{9}$$

where

$$K_3(s, t) = \int K(s, x) K_2(x, t) \, dx .\tag{10}$$

By continuing this process, and denoting

$$K_m(s, t) = \int K(s, x) K_{m-1}(x, t) \, dx ,\tag{11}$$

we get the $(n+1)$th approximate solution of integral equation (1) as

$$g_n(s) = f(s) + \sum_{m=1}^{n} \lambda^m \int K_m(s, t) f(t) \, dt .\tag{12}$$

We call the expression $K_m(s, t)$ the mth iterate, where $K_1(s, t) = K(s, t)$. Passing to the limit as $n \to \infty$, we obtain the so-called Neumann series

$$g(s) = \lim_{n \to \infty} g_n(s) = f(s) + \sum_{m=1}^{\infty} \lambda^m \int K_m(s, t) f(t) \, dt .\tag{13}$$

It remains to determine conditions under which this convergence is achieved. For this purpose, we attend to the partial sum (12) and apply the Schwarz inequality (1.6.3) to the general term of this sum. This gives

$$\left| \int K_m(s,t)f(t)\,dt \right|^2 \leqslant \left(\int |K_m(s,t)|^2\,dt \right) \int |f(t)|^2\,dt \ . \qquad (14)$$

Let D be the norm of f,

$$D^2 = \int |f(t)|^2\,dt \ , \qquad (15)$$

and let $C_m{}^2$ denote the upper bound of the integral

$$\int |K_m(s,t)|^2\,dt \ .$$

Hence, the inequality (14) becomes

$$\left| \int K_m(s,t)f(t)\,dt \right|^2 \leqslant C_m{}^2\,D^2 \ . \qquad (16)$$

The next step is to connect the estimate $C_m{}^2$ with the estimate $C_1{}^2$. This is achieved by applying the Schwarz inequality to the relation (11):

$$|K_m(s,t)|^2 \leqslant \int |K_{m-1}(s,x)|^2\,dx \int |K(x,t)|^2\,dx \ ,$$

which, when integrated with respect to t, yields

$$\int |K_m(s,t)|^2\,dt \leqslant B^2\,C_{m-1}^2 \ , \qquad (17)$$

where

$$B^2 = \iint |K(x,t)|^2\,dx\,dt \ . \qquad (18)$$

The inequality (17) sets up the recurrence relation

$$C_m{}^2 \leqslant B^{2m-2}\,C_1{}^2 \ . \qquad (19)$$

From (16) and (19), we have the inequality

$$\left| \int K_m(s,t)f(t)\,dt \right|^2 \leqslant C_1{}^2\,D^2\,B^{2m-2} \ . \qquad (20)$$

Therefore, the general term of the partial sum (12) has a magnitude less than the quantity $DC_1\,|\lambda|^m\,B^{m-1}$, and it follows that the infinite series (13) converges faster than the geometric series with common ratio $|\lambda|B$. Hence, if

$$|\lambda|\,B < 1 \ , \qquad (21)$$

the uniform convergence of this series is assured.

It will now be proved that, for given λ, equation (1) has a unique solution. Suppose the contrary, and let $g_1(s)$ and $g_2(s)$ be two solutions of equation (1):

$$g_1(s) = f(s) + \lambda \int K(s,t)g_1(t)\,dt\,,$$

$$g_2(s) = f(s) + \lambda \int K(s,t)g_2(t)\,dt\,.$$

By subtracting these equations and setting $g_1(s) - g_2(s) = \phi(s)$, there results the homogeneous integral equation

$$\phi(s) = \lambda \int K(s,t)\phi(t)\,dt\,.$$

Apply the Schwarz inequality to this equation and get

$$|\phi(s)|^2 \leqslant |\lambda|^2 \int |K(s,t)|^2\,dt \int |\phi(t)|^2\,dt\,,$$

which, when integrated with respect to s, becomes

$$\int |\phi(s)|^2\,ds \leqslant |\lambda|^2 \iint |K(s,t)|^2\,ds\,dt \int |\phi(s)|^2\,ds$$

or

$$(1 - |\lambda|^2\,B^2) \int |\phi(s)|^2\,ds \leqslant 0\,. \tag{22}$$

In view of the inequality (21) and the nature of the function $\phi(s) = g_1(s) - g_2(s)$, we readily conclude that $\phi(s) \equiv 0$, i.e., $g_1(s) \equiv g_2(s)$.

What is the estimate of the error for neglecting terms after the nth term in the Neumann series (13)? This is found by writing this series as

$$g(s) = f(s) + \sum_{m=1}^{n} \lambda^m \int K_m(s,t)f(t)\,dt + R_n(s)\,. \tag{23}$$

Then, it follows from the above analysis that

$$|R_n| \leqslant DC_1\,|\lambda|^{n+1}\,B^n/(1 - |\lambda|\,B)\,. \tag{24}$$

Finally, we can evaluate the resolvent kernel, as defined in the previous chapter, in terms of the iterated kernels $K_m(s,\,t)$. Indeed, by changing the order of integration and summation in the Neumann series (13), we obtain

$$g(s) = f(s) + \lambda \int \Big[\sum_{m=1}^{\infty} \lambda^{m-1} K_m(s,t)\Big] f(t)\,dt\,.$$

Comparing this with (2.3.7),

$$g(s) = f(s) + \lambda \int \Gamma(s,t;\lambda)f(t)\,dt\,, \tag{25}$$

we have

$$\Gamma(s,t;\lambda) = \sum_{m=1}^{\infty} \lambda^{m-1} K_m(s,t) . \tag{26}$$

From the above discussion, we can infer (see Section 3.5) that the series (26) is also convergent at least for $|\lambda| B < 1$. Hence, the resolvent kernel is an analytic function of λ, regular at least inside the circle $|\lambda| < B^{-1}$.

From the uniqueness of the solution of (1), we can prove that the resolvent kernel $\Gamma(s,t;\lambda)$ is unique. In fact, let equation (1) have, with $\lambda = \lambda_0$, two resolvent kernels $\Gamma_1(s,t;\lambda_0)$ and $\Gamma_2(s,t;\lambda_0)$. In view of the uniqueness of the solution of (1), an arbitrary function $f(s)$ satisfies the identity

$$f(s) + \lambda_0 \int \Gamma_1(s,t;\lambda_0)f(t)\,dt \equiv f(s) + \lambda_0 \int \Gamma_2(s,t;\lambda_0)f(t)\,dt . \tag{27}$$

Setting $\Psi(s,t;\lambda_0) = \Gamma_1(s,t;\lambda_0) - \Gamma_2(s,t;\lambda_0)$, we have, from (27),

$$\int \Psi(s,t;\lambda_0)f(t)\,dt \equiv 0 ,$$

for an arbitrary function $f(t)$. Let us take $f(t) = \Psi^*(s,t;\lambda)$, with fixed s. This implies that

$$\int |\Psi(s,t;\lambda_0)|^2\,dt \equiv 0 ,$$

which means that $\Psi(s,t;\lambda_0) \equiv 0$, proving the uniqueness of the resolvent kernel.

The above analysis can be summed up in the following basic theorem.

Theorem. To each $\mathscr{L}_2$-kernel $K(s,t)$, there corresponds a unique resolvent kernel $\Gamma(s,t;\lambda)$ which is an analytic function of λ, regular at least inside the circle $|\lambda| < B^{-1}$, and represented by the power series (26). Furthermore, if $f(s)$ is also an $\mathscr{L}_2$-function, then the unique, $\mathscr{L}_2$-solution of the Fredholm equation (1) valid in the circle $|\lambda| < B^{-1}$ is given by the formula (25).

The method of successive approximations has many drawbacks. In addition to being a cumbersome process, the Neumann series, in general, cannot be summed in closed form. Furthermore, a solution of the integral equation (1) may exist even if $|\lambda| B > 1$, as evidenced in the previous chapter. In fact, we saw that the resolvent kernel is a quotient of two polynomials of nth degree in λ, and therefore the only possible singular points of $\Gamma(s, t; \lambda)$ are the roots of the denominator $D(\lambda) = 0$. But, for $|\lambda| B > 1$, the Neumann series does not converge and as such does not provide the desired solution. We shall have more to say about these ideas in the next chapter.

3.2. EXAMPLES

Example 1.   Solve the integral equation

$$g(s) = f(s) + \lambda \int_0^1 e^{s-t} g(t) \, dt \, . \tag{1}$$

Following the method of the previous section, we have

$$K_1(s, t) = e^{s-t} \, ,$$

$$K_2(s, t) = \int_0^1 e^{s-x} e^{x-t} \, dx = e^{s-t} \, .$$

Proceeding in this way, we find that all the iterated kernels coincide with $K(s, t)$. Using (3.1.26), we obtain the resolvent kernel as

$$\Gamma(s, t; \lambda) = K(s, t)(1 + \lambda + \lambda^2 + \cdots) = e^{s-t}/(1 - \lambda) \, . \tag{2}$$

Although the series $(1 + \lambda + \lambda^2 + \cdots)$ converges only for $|\lambda| < 1$, the resolvent kernel is, in fact, an analytic function of λ, regular in the whole plane except at the point $\lambda = 1$, which is a simple pole of the kernel Γ. The solution $g(s)$ then follows from (3.1.25):

$$g(s) = f(s) - [\lambda/(\lambda - 1)] \int_0^1 e^{s-t} f(t) \, dt \, . \tag{3}$$

Example 2.   Solve the Fredholm integral equation

$$g(s) = 1 + \lambda \int_0^1 (1 - 3st) g(t) \, dt \tag{4}$$

and evaluate the resolvent kernel.

Starting with $g_0(s) = 1$, we have

$$g_1(s) = 1 + \lambda \int_0^1 (1-3st)\, dt = 1 + \lambda(1-\tfrac{3}{2}s) ,$$

$$g_2(s) = 1 + \lambda \int_0^1 (1-3st)[1+\lambda(1-\tfrac{3}{2}t)]\, dt = 1 + \lambda(1-\tfrac{3}{2}s) + \tfrac{1}{4}\lambda^2 ,$$

$\cdots$

$$g(s) = 1 + \lambda(1-\tfrac{3}{2}s) + \tfrac{1}{4}\lambda^2 + \tfrac{1}{4}\lambda^3(1-\tfrac{3}{2}s) + \tfrac{1}{16}\lambda^4 + \tfrac{1}{16}\lambda^5(1-\tfrac{3}{2}s) + \cdots ,$$

or

$$g(s) = (1+\tfrac{1}{4}\lambda^2+\tfrac{1}{16}\lambda^4+\cdots)[1 + \lambda(1-\tfrac{3}{2}s)] . \tag{5}$$

The geometric series in (4) is convergent provided $|\lambda| < 2$. Then,

$$g(s) = [4 + 2\lambda(2-3s)]/(4-\lambda^2) , \tag{6}$$

and precisely the same remarks apply to the region of the validity of this solution as given in Example 1.

To evaluate the resolvent kernel, we find the iterated kernels

$$K_1(s,t) = 1 - 3st ,$$

$$K_2(s,t) = \int_0^1 (1-3sx)(1-3xt)\, dx = 1 - \tfrac{3}{2}(s+t) + 3st ,$$

$$K_3(s,t) = \int_0^1 (1-3sx)[1 - \tfrac{3}{2}(x+t) - 3xt]\, dx$$

$$= \tfrac{1}{4}(1-3st) = \tfrac{1}{4}K_1(s,t) ,$$

Similarly,

$$K_4(s,t) = \tfrac{1}{4}K_2(s,t)$$

and

$$K_n(s,t) = \tfrac{1}{4}K_{n-2}(s,t) .$$

Hence,

$$\Gamma(s,t;\lambda) = K_1 + \lambda K_2 + \lambda^2 K_3 + \cdots$$

$$= (1+\tfrac{1}{4}\lambda^2+\tfrac{1}{16}\lambda^4+\cdots)K_1 + \lambda(1+\tfrac{1}{4}\lambda^2+\tfrac{1}{16}\lambda^4+\cdots)K_2$$

$$= [(1+\lambda) - \tfrac{3}{2}\lambda(s+t) - 3(1-\lambda)st]/(1-\tfrac{1}{4}\lambda^2) , \tag{7}$$

$$|\lambda| < 2 .$$

Example 3.   Solve the integral equation

$$g(s) = 1 + \lambda \int_0^\pi [\sin(s+t)] g(t) \, dt \, . \tag{8}$$

Let us first evaluate the iterated kernels in this example:

$$K_1(s,t) = K(s,t) = \sin(s+t) \, ,$$

$$K_2(s,t) = \int_0^\pi [\sin(s+x)] \sin(x+t) \, dx$$

$$= \tfrac{1}{2}\pi [\sin s \sin t + \cos s \cos t] = \tfrac{1}{2}\pi \cos(s-t) \, ,$$

$$K_3(s,t) = \tfrac{1}{2}\pi \int_0^1 [\sin(s+x)] \cos(x-t) \, dx$$

$$= \tfrac{1}{2}\pi \int_0^1 (\sin s \cos x + \sin x \cos s)$$

$$\times (\cos x \cos t + \sin x \sin t) \, dx$$

$$= (\tfrac{1}{2}\pi)^2 [\sin s \cos t + \cos s \sin t] = (\tfrac{1}{2}\pi)^2 \sin(s+t) \, .$$

Proceeding in this manner, we obtain

$$K_4(s,t) = (\tfrac{1}{2}\pi)^3 \cos(s-t) \, ,$$
$$K_5(s,t) = (\tfrac{1}{2}\pi)^4 \sin(s+t) \, ,$$
$$K_6(s,t) = (\tfrac{1}{2}\pi)^5 \cos(s-t) \, , \quad \text{etc.}$$

Substituting these values in the formula (3.1.13) and integrating, there results the solution

$$g(s) = 2\lambda(\cos s) [1 + (\tfrac{1}{2}\pi)^2 \lambda^2 + (\tfrac{1}{2}\pi)^4 \lambda^4 + \cdots]$$
$$+ \lambda^2 \pi (\sin s) [1 + (\tfrac{1}{2}\pi)^2 \lambda^2 + (\tfrac{1}{2}\pi)^4 \lambda^4 + \cdots] \, , \tag{9}$$

or

$$g(s) = 1 + [(2\lambda \cos s + \lambda^2 \pi \sin s)/(1 - \tfrac{1}{4}\lambda^2 \pi^2)] \, . \tag{10}$$

Since

$$B^2 = \int_0^\pi \int_0^\pi \sin^2(s+t) \, ds \, dt = \tfrac{1}{2}\pi^2 \, ,$$

the interval of convergence of the series (9) lies between $-\sqrt{2}/\pi$ and $\sqrt{2}/\pi$.

Example 4.   Prove that the mth iterated kernel $K_m(s, t)$ satisfies the following relation:

$$K_m(s, t) = \int K_r(s, x) K_{m-r}(x, t) \, dx \, , \tag{11}$$

where r is any positive integer less than n.

By successive applications of (3.1.11),

$$K_m(s, t) = \int \cdots \int K(s, x_1) K(x_1, x_2) \cdots K(x_{m-1}, t) \, dx_{m-1} \cdots dx_1 \, . \tag{12}$$

Thus, $K_m(s, t)$ is an $(m-1)$-fold integral. Similarly, $K_r(s, x)$ and $K_{m-r}(x, t)$ are $(r-1)$- and $(m-r-1)$-fold integrals. This means that

$$\int K_r(s, x) K_{m-r}(x, t) \, dx$$

is an $(m-1)$-fold integral, and the result follows.

One can take the $g_0(s)$ approximation different from $f(s)$, as we demonstrate by the following example.

Example 5.   Solve the inhomogeneous Fredholm integral equation of the second kind,

$$g(s) = 2s + \lambda \int_0^1 (s+t) g(t) \, dt \, , \tag{13}$$

by the method of successive approximations to the third order.

For this equation, we take $g_0(s) = 1$. Then,

$$g(s) = 2s + \lambda \int_0^1 (s+t) \, dt = 2s + \lambda(s+\tfrac{1}{2}) \, ,$$

$$g_2(s) = 2s + \lambda \int_0^1 (s+t) \{2t + \lambda[t+(1/2)]\} \, dt$$

$$= 2s + \lambda[s+(2/3)] + \lambda^2[s+(7/12)] \, ,$$

$$g_3(s) = 2s + \lambda \int_0^1 (s+t) \{2t + \lambda[t+(2/3)] + \lambda^2[t+(7/12)]\} \, dt$$

$$= 2s + \lambda[s+(2/3)] + \lambda^2[(7/6)s+(2/3)] + \lambda^3[(13/12)s+(5/8)] \, .$$

$$\tag{14}$$

From Example 2 of Section 2.2, we find the exact solution to be

$$g(s) = [12(2-\lambda)s + 8\lambda]/(12-2\lambda-\lambda^2) , \qquad (15)$$

and the comparison of (14) and (15) is left to the reader.

3.3. VOLTERRA INTEGRAL EQUATION

The same iterative scheme is applicable to the Volterra integral equation of the second kind. In fact, the formulas corresponding to (3.1.13) and (3.1.25) are, respectively,

$$g(s) = f(s) + \sum_{m=1}^{\infty} \lambda^m \int_a^s K_m(s, t) f(t) \, dt , \qquad (1)$$

$$g(s) = f(s) + \lambda \int_a^s \Gamma(s, t; \lambda) f(t) \, dt , \qquad (2)$$

where the iterated kernel $K_m(s, t)$ satisfies the recurrence formula

$$K_m(s, t) = \int_t^s K(s, x) K_{m-1}(x, t) \, dx \qquad (3)$$

with $K_1(s, t) = K(s, t)$, as before. The resolvent kernel $\Gamma(s, t; \lambda)$ is given by the same formula as (3.1.26), and it is an entire function of λ for any given (s, t) (see Exercise 8).

We shall illustrate it by the following examples.

3.4. EXAMPLES

Example 1. Find the Neumann series for the solution of the integral equation

$$g(s) = (1+s) + \lambda \int_0^s (s-t) g(t) \, dt . \qquad (1)$$

From the formula (3.3.3), we have

$$K_1(s,t) = (s-t),$$

$$K_2(s,t) = \int_t^s (s-x)(x-t)\,dx = \frac{(s-t)^3}{3!},$$

$$K_3(s,t) = \int_t^s \frac{(s-x)(x-t)^3}{3!}\,dx = \frac{(s-t)^5}{5!},$$

and so on. Thus,

$$g(s) = 1 + s + \lambda\left(\frac{s^2}{2!} + \frac{s^3}{3!}\right) + \lambda^2\left(\frac{s^4}{4!} + \frac{s^5}{5!}\right) + \cdots. \tag{2}$$

For $\lambda = 1$, $g(s) = e^s$.

Example 2. Solve the integral equation

$$g(s) = f(s) + \lambda \int_0^s e^{s-t} g(t)\,dt \tag{3}$$

and evaluate the resolvant kernel.

For this case,

$$K_1(s,t) = e^{s-t},$$

$$K_2(s,t) = \int_t^s e^{s-x} e^{x-t}\,dx = (s-t)e^{s-t},$$

$$K_3(s,t) = \int_t^s (x-t)e^{s-x} e^{x-t}\,dx = \frac{(s-t)^2}{2!}e^{s-t},$$

$$K_m(s,t) = \frac{(s-t)^{m-1}}{(m-1)!}e^{s-t}.$$

The resolvent kernel is

$$\Gamma(s,t;\lambda) = \begin{cases} e^{s-t}\displaystyle\sum_{m=1}^{\infty} \frac{\lambda^{m-1}(s-t)^{m-1}}{(m-1)!} = e^{(\lambda+1)(s-t)}, & t \leqslant s, \\[2mm] 0, & t > s. \end{cases} \tag{4}$$

Hence, the solution is

$$g(s) = f(s) + \lambda \int_0^s e^{(\lambda+1)(s-t)} f(t) \, dt \, . \tag{5}$$

Example 3.   Solve the Volterra equation

$$g(s) = 1 + \int_0^s stg(t) \, dt \, . \tag{6}$$

For this example, $K_1(s, t) = K(s, t) = st$,

$$K_2(s, t) = \int_t^s sx^2 t \, dx = (s^4 t - st^4)/3 \, ,$$

$$K_3(s, t) = \int_t^s [(sx)(x^4 t - xt^4)/3] \, dx = (s^7 t - 2s^4 t^4 + st^7)/18 \, ,$$

$$K_4(s, t) = \int_t^s [(sx)(x^7 t - 2x^4 t^4 + xt^7)/18] \, dx$$

$$= (s^{10} t - 3s^7 t + 3s^4 t^7 - st^{10})/162 \, ,$$

and so on. Thus.

$$g(s) = 1 + \frac{s^3}{3} + \frac{s^6}{2 \cdot 5} + \frac{s^9}{2 \cdot 5 \cdot 8} + \frac{s^{12}}{2 \cdot 5 \cdot 8 \cdot 11} + \cdots \, . \tag{7}$$

3.5. SOME RESULTS ABOUT THE RESOLVENT KERNEL

The series for the resolvent kernel $\Gamma(s, t; \lambda)$,

$$\Gamma(s, t; \lambda) = \sum_{m=1}^{\infty} \lambda^{m-1} K_m(s, t) \, , \tag{1}$$

can be proved to be absolutely and uniformly convergent for all values of s and t in the circle $|\lambda| < 1/B$. In addition to the assumptions of Section 3.1, we need the additional inequality

$$\int |K(s, t)|^2 \, ds < E^2 \, , \qquad E = \text{const} \, . \tag{2}$$

Recall that this is one of the conditions for the kernel K to be an $\mathscr{L}_2$-kernel. Applying the Schwarz inequality to the recurrence formula

$$K_m(s, t) = \int K_{m-1}(s, x) K(x, t)\, dx \tag{3}$$

yields

$$|K_m(s, t)|^2 \leqslant \left(\int |K_{m-1}(s, x)|^2\, dx \right) \int |K(x, t)|^2\, dx\ ,$$

which, with the help of (3.1.19), becomes

$$|K_m(s, t)| \leqslant C_1\, EB^{m-1}\ . \tag{4}$$

Thus, the series (1) is dominated by the geometric series with the general term $C_1\, E(\lambda^{m-1}\, B^{m-1})$, and that completes the proof.

Next, we prove that the resolvent kernel satisfies the integral equation

$$\Gamma(s, t; \lambda) = K(s, t) + \lambda \int K(s, x)\, \Gamma(x, t; \lambda)\, dx\ . \tag{5}$$

This follows by replacing $K_m(s, t)$ in the series (1) by the integral relation (3). Then,

$$\Gamma(s, t; \lambda) = K_1(s, t) + \sum_{m=2}^{\infty} \lambda^{m-1} \int K_{m-1}(s, x) K(x, t)\, dx$$

$$= K(s, t) + \lambda \sum_{m=1}^{\infty} \lambda^{m-1} \int K_m(s, x) K(x, t)\, dx$$

$$= K(s, t) + \lambda \int \left[\sum_{m=1}^{\infty} \lambda^{m-1} K_m(s, x) \right] K(x, t)\, dx\ ,$$

and the integral equation (5) follows immediately. The change of order of integration and summation is legitimate in view of the uniform convergence of the series involved.

Another interesting result for the resolvent kernel is that it satisfies the integrodifferential equation

$$\partial \Gamma(s, t; \lambda)/\partial \lambda = \int \Gamma(s, x; \lambda)\, \Gamma(x, t; \lambda)\, dx\ . \tag{6}$$

In fact,

$$\int \Gamma(s, x; \lambda)\, \Gamma(x, t; \lambda)\, dx = \int \sum_{m=1}^{\infty} \lambda^{m-1} K_m(s, x) \sum_{n=1}^{\infty} \lambda^{n-1} K_n(x, t)\, dx\ .$$

On account of the absolute and uniform convergence of the series (1), we can multiply the series under the integral sign and integrate it term by term. Therefore,

$$\int \Gamma(s, x; \lambda) \Gamma(x, t; \lambda) \, dx = \sum_{m=1}^{\infty} \sum_{n=1}^{\infty} \lambda^{m+n-2} K_{m+n}(s, t) \, . \tag{7}$$

Now, set $m+n=p$ and change the order of summation; there results the relation

$$\sum_{m=1}^{\infty} \sum_{n=1}^{\infty} \lambda^{m+n-2} K_{m+n}(s, t) = \sum_{p=2}^{\infty} \sum_{n=1}^{p-1} \lambda^{p-2} K_p(s, t)$$

$$= \sum_{p=2}^{\infty} (p-1) \lambda^{p-2} K_p(s, t) = \frac{\partial \Gamma(s, t; \lambda)}{\partial \lambda} \, . \tag{8}$$

Combining (7) and (8), we have the result (6).

EXERCISES

1. Solve the following Fredholm integral equations by the method of successive approximations:

$$\text{(i)} \quad g(s) = e^s - \tfrac{1}{2}e + \tfrac{1}{2} + \tfrac{1}{2} \int_0^1 g(t) \, dt \, .$$

$$\text{(ii)} \quad g(s) = (\sin s) - \tfrac{1}{4}s + \tfrac{1}{4} \int_0^{\pi/2} stg(t) \, dt \, .$$

2. Consider the integral equation

$$g(s) = 1 + \lambda \int_0^1 stg(t) \, dt \, .$$

(a) Make use of the relation $|\lambda| < B^{-1}$ to show that the iterative procedure is valid for $|\lambda| < 3$.

(b) Show that the iterative procedure leads formally to the solution

$$g(s) = 1 + s[(\lambda/2) + (\lambda^2/6) + (\lambda^3/18) + \cdots] \, .$$

(c) Use the method of the previous chapter to obtain the exact solution

$$g(s) = 1 + [3\lambda s/2(3-\lambda)] \, , \qquad \lambda \neq 3 \, .$$

3. Solve the integral equation

$$g(s) = 1 + \lambda \int_0^1 (s+t) g(t) \, dt \, ,$$

by the method of successive approximations and show that the estimate afforded by the relation $|\lambda| < B^{-1}$ is conservative in this case.

4. Find the resolvent kernel associated with the following kernels: (i) $|s-t|$, in the interval $(0, 1)$; (ii) $\exp - |s-t|$, in the interval $(0, 1)$; (iii) $\cos(s+t)$, in the interval $(0, 2\pi)$.

5. Solve the following Volterra integral equations by the method of this chapter:

$$(i) \quad g(s) = 1 + \int_0^s (s-t) g(t) \, dt \, ,$$

$$(ii) \quad g(s) = 29 + 6s + \int_0^s (6s - 6t + 5) g(t) \, dt \, .$$

6. Find an approximate solution of the integral equation

$$g(s) = (\sinh s) + \int_0^s e^{t-s} g(t) \, dt \, ,$$

by the method of iteration.

7. Obtain the radius of convergence of the Neumann series when the functions $f(s)$ and the kernel $K(s, t)$ are continuous in the interval (a, b).

8. Prove that the resolvent kernel for a Volterra integral equation of the second kind is an entire function of λ for any given (s, t).

CLASSICAL FREDHOLM THEORY

4.1. THE METHOD OF SOLUTION OF FREDHOLM

In the previous chapter, we have derived the solution of the Fredholm integral equation

$$g(s) = f(s) + \lambda \int K(s, t) g(t) \, dt \tag{1}$$

as a uniformly convergent power series in the parameter λ for $|\lambda|$ suitably small. Fredholm gave the solution of equation (1) in general form for all values of the parameter λ. His results are contained in three theorems which bear his name. We have already studied them in Chapter 2 for the special case when the kernel is separable. In this chapter, we shall study equation (1) when the function $f(s)$ and the kernel $K(s, t)$ are any integrable functions. Furthermore, the present method enables us to get explicit formulas for the solution in terms of certain determinants.

The method used by Fredholm consists in viewing the integral equation (1) as the limiting case of a system of linear algebraic equations. This theory applies to two- or higher-dimensional integrals, although we shall confine our discussion to only one-dimensional integrals in the interval (a, b). Let us divide the interval (a, b) into n equal parts,

$$s_1 = t_1 = a, \quad s_2 = t_2 = a + h, \quad \dots, \quad s_n = t_n = a + (n-1)h,$$

where $h = (b-a)/n$. Thereby, we have the approximate formula

$$\int K(s,t)g(t)\,dt \simeq h\sum_{j=1}^{n} K(s,s_j)g(s_j)\,. \tag{2}$$

Equation (1) then takes the form

$$g(s) \simeq f(s) + \lambda h\sum_{j=1}^{n} K(s,s_j)g(s_j)\,, \tag{3}$$

which must hold for all values of s in the interval (a,b). In particular, this equation is satisfied at the n points of division s_i, $i = 1, \ldots, n$. This leads to the system of equations

$$g(s_i) = f(s_i) + \lambda h\sum_{j=1}^{n} K(s_i,s_j)g(s_j)\,, \qquad i = 1,\ldots,n\,. \tag{4}$$

Writing

$$f(s_i) = f_i\,, \qquad g(s_i) = g_i\,, \qquad K(s_i,s_j) = K_{ij}\,, \tag{5}$$

equation (4) yields an approximation for the integral equation (1) in terms of the system of n linear equations

$$g_i - \lambda h\sum_{j=1}^{n} K_{ij}g_j = f_i\,, \qquad i = 1,\ldots,n\,, \tag{6}$$

in n unknown quantities $g_1, \ldots, g_n$. The values of g_i obtained by solving this algebraic system are approximate solutions of the integral equation (1) at the points $s_1, s_2, \ldots, s_n$. We can plot these solutions g_i as ordinates and by interpolation draw a curve $g(s)$ which we may expect to be an approximation to the actual solution. With the help of this algebraic system, we can also determine approximations for the eigenvalues of the kernel.

The resolvent determinant of the algebraic system (6) is

$$D_n(\lambda) = \begin{vmatrix} 1-\lambda h K_{11} & -\lambda h K_{12} & \cdots & -\lambda h K_{1n} \\ -\lambda h K_{21} & 1-\lambda h K_{22} & \cdots & -\lambda h K_{2n} \\ \vdots & & & \\ -\lambda h K_{n1} & -\lambda h K_{n2} & \cdots & 1-\lambda h K_{nn} \end{vmatrix}. \tag{7}$$

The approximate eigenvalues are obtained by setting this determinant equal to zero. We illustrate it by the following example.

Example.

$$g(s) - \lambda \int_0^{\pi} \sin(s+t)\,g(t)\,dt = 0 .$$

By taking $n = 3$, we have $h = \pi/3$ and therefore

$$s_1 = t_1 = 0 , \qquad s_2 = t_2 = \pi/3 , \qquad s_3 = t_3 = 2\pi/3 ,$$

and the values of K_{ij} are readily calculated to give

$$(K_{ij}) = \begin{vmatrix} 0 & 0.866 & 0.866 \\ 0.866 & 0.866 & 0 \\ 0.866 & 0 & -0.866 \end{vmatrix} .$$

The homogeneous system corresponding to (6) will have a non-trivial solution if the determinant

$$D_n(\lambda) = \begin{vmatrix} 1 & -0.907\lambda & -0.907\lambda \\ -0.907\lambda & (1-0.907\lambda) & 0 \\ -0.907\lambda & 0 & (1+0.907\lambda) \end{vmatrix} = 0 ,$$

or when $1 - 3(0.0907)^2 \lambda^2 = 0$. The roots of this equation are $\lambda = \pm 0.6365$. This gives a rather close agreement with the exact values (see Example 3, Section 3.2), which are $\pm \sqrt{2}/\pi = \pm 0.6366$.

In general, the practical applications of this method are limited because one has to take a rather large n to get a reasonable approximation.

4.2. FREDHOLM'S FIRST THEOREM

The solutions $g_1, g_2, ..., g_n$ of the system of equations (4.1.6) are obtained as ratios of certain determinants, with the determinant $D_n(\lambda)$ given by (4.1.7) as the denominator provided it does not vanish. Let us expand the determinant (4.1.7) in powers of the quantity $(-\lambda h)$. The constant term is obviously equal to unity. The term containing $(-\lambda h)$ in the first power is the sum of all the determinants containing only one column $-\lambda h K_{\mu\nu}$, $\mu = 1, \ldots, n$. Taking the contribution from all the columns $\nu = 1, \ldots, n$, we find that the total contribution is $-\lambda h \sum_{\nu=1}^{n} K_{\nu\nu}$.

The factor containing the factor $(-\lambda h)$ to the second power is the sum of all the determinants containing two columns with that factor. This results in the determinants of the form

$$(-\lambda h)^2 \begin{vmatrix} K_{pp} & K_{pq} \\ K_{qp} & K_{qq} \end{vmatrix},$$

where (p, q) is an arbitrary pair of integers taken from the sequence $1, \ldots, n$, with $p < q$. In the same way, it follows that the term containing the factor $(-\lambda h)^3$ is the sum of the determinants of the form

$$(-\lambda h)^3 \begin{vmatrix} K_{pp} & K_{pq} & K_{pr} \\ K_{qp} & K_{qq} & K_{qr} \\ K_{rp} & K_{rq} & K_{rr} \end{vmatrix},$$

where (p, q, r) is an arbitrary triplet of integers selected from the sequence $1, \ldots, n$, with $p < q < r$.

The remaining terms are obtained in a similar manner. Therefore, we conclude that the required expansion of $D_n(\lambda)$ is

$$D_n(\lambda) = 1 - \lambda h \sum_{\nu=1}^{n} K_{\nu\nu} + \frac{(-\lambda h)^2}{2!} \sum_{p,q=1}^{n} \begin{vmatrix} K_{pp} & K_{pq} \\ K_{qp} & K_{qq} \end{vmatrix}$$

$$+ \frac{(-\lambda h)^3}{3!} \sum_{p,q,r=1}^{n} \begin{vmatrix} K_{pp} & K_{pq} & K_{pr} \\ K_{qp} & K_{qq} & K_{qr} \\ K_{rp} & K_{rq} & K_{rr} \end{vmatrix} + \cdots$$

$$+ \frac{(-\lambda h)^n}{n!} \sum_{p_1, p_2, \ldots, p_n = 1}^{n} \begin{vmatrix} K_{p_1 p_1} & K_{p_1 p_2} & \cdots & K_{p_1 p_n} \\ K_{p_2 p_1} & K_{p_2 p_2} & \cdots & K_{p_2 p_n} \\ \vdots & & & \\ K_{p_n p_1} & K_{p_n p_2} & \cdots & K_{p_n p_n} \end{vmatrix}, \qquad (1)$$

where we now stipulate that the sums are taken over all permutations of pairs (p, q), triplets (p, q, r), etc. This convention explains the reason for dividing each term of the above series by the corresponding number of permutations.

The analysis is simplified by introducing the following symbol for the determinant formed by the values of the kernel at all points (s_i, t_j)

$$\begin{vmatrix} K(s_1,t_1) & K(s_1,t_2) & \cdots & K(s_1,t_n) \\ K(s_2,t_1) & K(s_2,t_2) & \cdots & K(s_2,t_n) \\ \vdots & & & \\ K(s_n,\ t_1) & K(s_n,\ t_2) & \cdots & K(s_n,\ t_n) \end{vmatrix} = K\begin{pmatrix} s_1,s_2,\ldots,s_n \\ t_1,t_2,\ldots,t_n \end{pmatrix}, \qquad (2)$$

the so-called Fredholm determinant. We observe that, if any pair of arguments in the upper or lower sequence is transposed, the value of the determinant changes sign because the transposition of two arguments in the upper sequence corresponds to the transposition of two rows of the determinant and the transposition of two arguments in the lower sequence corresponds to the transposition of two columns.

In this notation, the series (1) takes the form

$$D_n(\lambda) = 1 - \lambda h \sum_{p=1}^{n} K(s_p, s_p) + \frac{(-\lambda h)^2}{2!} \sum_{p,q=1}^{n} K\begin{pmatrix} s_p, s_q \\ s_p, s_q \end{pmatrix}$$

$$+ \frac{(-\lambda h)^3}{3!} \sum_{p,q,r=1}^{n} K\begin{pmatrix} s_p, s_q, s_r \\ s_p, s_q, s_r \end{pmatrix} + \cdots . \qquad (3)$$

If we now let n tend to infinity, then h will tend to zero, and each term of the sum (3) tends to some single, double, triple integral, etc. There results Fredholm's first series:

$$D(\lambda) = 1 - \lambda \int K(s,s)\, ds + \frac{\lambda^2}{2!} \int K\begin{pmatrix} s_1, s_2 \\ s_1, s_2 \end{pmatrix} ds_1\, ds_2$$

$$- \frac{\lambda^3}{3!} \iiint K\begin{pmatrix} s_1, s_2, s_3 \\ s_1, s_2, s_3 \end{pmatrix} ds_1\, ds_2\, ds_3 + \cdots . \qquad (4)$$

Hilbert gave a rigorous proof of the fact that the sequence $D_n(\lambda) \to D(\lambda)$ in the limit, while the convergence of the series (4) for all values of λ was proved by Fredholm on the basis that the kernel $K(s, t)$ is a bounded and integrable function.[1] Thus, $D(\lambda)$ is an entire function of the complex variable λ.

We are now ready to solve the Fredholm equation (4.1.1) and express

---

[1] For proof, see Lovitt [10].

the solutions in the form of a quotient of two power series in the parameter λ, where the Fredholm function $D(\lambda)$ is to be the divisor. In this connection, recall the relations (2.3.6) and (2.3.7). Indeed, we seek solutions of the form

$$g(s) = f(s) + \lambda \int \Gamma(s, t; \lambda) f(t) \, dt \, , \tag{5}$$

and expect the resolvent kernel $\Gamma(s, t; \lambda)$ to be the quotient

$$\Gamma(s, t; \lambda) = D(s, t; \lambda)/D(\lambda) \, , \tag{6}$$

where $D(s, t; \lambda)$, still to be determined, is the sum of certain functional series.

Now, we have proved in Section 3.5 that the resolvent $\Gamma(s, t; \lambda)$ itself satisfies a Fredholm integral equation of the second kind (3.5.5):

$$\Gamma(s, t; \lambda) = K(s, t) + \lambda \int K(s, x) \Gamma(x, t; \lambda) \, dx \, . \tag{7}$$

From (6) and (7), it follows that

$$D(s, t; \lambda) = K(s, t) D(\lambda) + \lambda \int K(s, x) D(x, t; \lambda) \, dx \, . \tag{8}$$

The form of the series (4) for $D(\lambda)$ suggests that we seek the solution of equation (8) in the form of a power series in the parameter λ:

$$D(s, t; \lambda) = C_0(s, t) + \sum_{p=1}^{\infty} \frac{(-\lambda)^p}{p!} C_p(s, t) \, . \tag{9}$$

For this purpose, write the numerical series (4) as

$$D(\lambda) = 1 + \sum_{p=1}^{\infty} \frac{(-\lambda)^p}{p!} c_p \, , \tag{10}$$

where

$$c_p = \int \cdots \int K \begin{pmatrix} s_1, s_2, \ldots, s_p \\ s_1, s_2, \ldots, s_p \end{pmatrix} ds_1 \cdots ds_p \, . \tag{11}$$

The next step is to substitute the series for $D(s, t; \lambda)$ and $D(\lambda)$ from (9) and (10) in (8) and compare the coefficients of equal powers of λ. The following relations result:

$$C_0(s, t) = K(s, t) \, , \tag{12}$$

$$C_p(s, t) = c_p K(s, t) - p \int K(s, x) C_{p-1}(x, t) \, dx \, . \tag{13}$$

Our contention is that we can write the function $C_p(s, t)$ in terms of the Fredholm determinant (2) in the following way:

$$C_p(s,t) = \int \cdots \int K\begin{pmatrix} s, x_1, x_2, \ldots, x_p \\ t, x_1, x_2, \ldots, x_p \end{pmatrix} dx_1 \cdots dx_p . \qquad (14)$$

In fact, for $p = 1$, the relation (13) becomes

$$
\begin{aligned}
C_1(s,t) &= c_1 K(s\ t) - \int K(s,x) C_0(x,t)\, dx \\
&= K(s,t) \int K(x,x)\, dx - \int K(s,x) K(x,t)\, dx \\
&= \int K\begin{pmatrix} s & x \\ t & x \end{pmatrix} dx ,
\end{aligned}
\qquad (15)
$$

where we have used (11) and (12).

To prove that (14) holds for general p, we expand the determinant under the integral sign in the relation:

$$K\begin{pmatrix} s, x_1, \ldots, x_p \\ t, x_1, \ldots, x_p \end{pmatrix} = \begin{vmatrix} K(s,t) & K(s,x_1) & \cdots & K(s,x_p) \\ K(x_1,t) & K(x_1,x_1) & \cdots & K(x_1,x_p) \\ \vdots & & & \\ K(x_p,t) & K(x_p,x_1) & \cdots & K(x_p,x_p) \end{vmatrix} ,$$

with respect to the elements of the given row, transposing in turn the first column one place to the right, integrating both sides, and using the definition of c_p as in (11); the required result then follows by induction.

From (9), (12), and (14) we derive Fredholm's second series:

$$D(s,t;\lambda) = K(s,t) + \sum_{p=1}^{\infty} \frac{(-\lambda)^p}{p!} \int \cdots \int K\begin{pmatrix} s, x_1, \ldots, x_p \\ t, x_1, \ldots, x_p \end{pmatrix} dx_1 \cdots dx_p . \quad (16)$$

This series also converges for all values of the parameter λ. It is interesting to observe the similarity between the series (4) and (16).

Having found both terms of the quotient (6), we have established the existence of a solution to the integral equation (4.1.1) for a bounded and integrable kernel $K(s,t)$, provided, of course, that $D(\lambda) \neq 0$. Since both terms of this quotient are entire functions of the parameter λ, it follows that the resolvent kernel $\Gamma(s,t;\lambda)$ is a meromorphic function of λ, i.e., an analytic function whose singularities may only be the poles, which in the present case are zeros of the divisor $D(\lambda)$.

Next, we prove that the solution in the form obtained by Fredholm is unique and is given by

$$g(s) = f(s) + \lambda \int \Gamma(s,t;\lambda) f(t)\, dt . \qquad (17)$$

In this connection, we first observe that the integral equation (7) satisfied by $\Gamma(s,t;\lambda)$ is valid for all values of λ for which $D(\lambda) \neq 0$. Indeed, (7) is known to hold for $|\lambda| < B^{-1}$ from the analysis of Chapter 3, and since both sides of this equation are now proved to be meromorphic, the above contention follows. To prove the uniqueness of the solution, let us suppose that $g(s)$ is a solution of the equation (4.1.1) in the case $D(\lambda) \neq 0$. Multiply both sides of (4.1.1) by $\Gamma(s,t;\lambda)$, integrate, and get

$$\int \Gamma(s,x;\lambda)g(x)\,dx = \int \Gamma(s,x;\lambda)f(x)\,dx$$
$$+ \lambda \int \left[\int \Gamma(s,x;\lambda)K(x,t)\,dx \right] g(t)\,dt. \quad (18)$$

Substituting from (7) into left side of (18), this becomes

$$\int K(s,t)g(t)\,dt = \int \Gamma(s,x;\lambda)f(x)\,dx, \quad (19)$$

which, when joined by (4.1.1), yields

$$g(s) = f(s) + \lambda \int \Gamma(s,t;\lambda)f(t)\,dt, \quad (20)$$

but this form is unique.

In particular, the solution of the homogeneous equation

$$g(s) = \lambda \int K(s,t)g(t)\,dt \quad (21)$$

is identically zero.

The above analysis leads to the following theorem.

Fredholm's First Theorem. The inhomogeneous Fredholm equation

$$g(s) = f(s) + \lambda \int K(s,t)g(t)\,dt, \quad (22)$$

where the functions $f(s)$ and $g(t)$ are integrable, has a unique solution

$$g(s) = f(s) + \lambda \int \Gamma(s,t;\lambda)f(t)\,dt, \quad (23)$$

where the resolvent kernel $\Gamma(s,t;\lambda)$,

$$\Gamma(s,t;\lambda) = D(s,t;\lambda)/D(\lambda), \quad (24)$$

with $D(\lambda) \neq 0$, is a meromorphic function of the complex variable λ, being the ratio of two entire functions defined by the series

$$D(s,t;\lambda) = K(s,t) + \sum_{p=1}^{\infty} \frac{(-\lambda)^p}{p!} \int \cdots \int K\begin{pmatrix} s, x_1, \ldots, x_p \\ t, x_1, \ldots, x_p \end{pmatrix} dx_1 \cdots dx_p, \quad (25)$$

and

$$D(\lambda) = 1 + \sum_{p=1}^{\infty} \frac{(-\lambda)^p}{p!} \int \cdots \int K\begin{pmatrix} x_1, \ldots, x_p \\ x_1, \ldots, x_p \end{pmatrix} dx_1 \cdots dx_p , \qquad (26)$$

both of which converge for all values of λ. In particular, the solution of the homogeneous equation

$$g(s) = \lambda \int K(s,t) g(t) dt \qquad (27)$$

is identically zero.

4.3. EXAMPLES

Example 1. Evaluate the resolvent for the integral equation

$$g(s) = f(s) + \lambda \int_0^1 (s+t) g(t) dt . \qquad (1)$$

The solution to this example is obtained by writing

$$\Gamma(s,t;\lambda) = \left[\sum_{p=0}^{\infty} \frac{(-\lambda)^p}{p!} C_p(s,t) \right] \Big/ \sum_{p=0}^{\infty} \frac{(-\lambda)^p}{p!} c_p , \qquad (2)$$

where C_p and c_p are defined by the relations (4.2.11) and (4.2.13):

$$c_0 = 1 , \qquad C_0(s,t) = K(s,t) = (s+t) . \qquad (3)$$

$$c_p = \int C_{p-1}(s,s) ds , \qquad (4)$$

$$C_p = c_p K(s,t) - p \int_0^1 K(s,x) c_{p-1}(x,t) dx . \qquad (5)$$

Thus,

$$c_1 = \int_0^1 2s \, ds = 1 ,$$

$$C_1(s,t) = (s+t) - \int_0^1 (s+x)(x+t) \, dx = \tfrac{1}{2}(s+t) - st - \tfrac{1}{3} ,$$

$$c_2 = \int_0^1 (s-s^2-\tfrac{1}{3}) \, ds = -\tfrac{1}{6} ,$$

$$C_2(s,t) = -\tfrac{1}{6}(s+t) - 2\int_0^1 (s+x)[\tfrac{1}{2}(x+t) - xt - \tfrac{1}{3}])\, dx = 0 \;.$$

Since $C_2(x,t)$ vanishes, it follows from (5) that the subsequent coefficients C_k and c_k also vanish. Therefore,

$$\Gamma(s,t;\lambda) = \frac{(s+t) - [\tfrac{1}{2}(s+t) - st - \tfrac{1}{3}]\lambda}{1 - \lambda - (\lambda^2/12)}\;, \tag{6}$$

which agrees with result (2.2.8) found by a different method.

Example 2.   Solve the integral equation

$$g(s) = s + \lambda \int_0^1 [st + (st)^{1/2}]\, g(t)\, dt \;. \tag{7}$$

In this case,

$$c_0 = 1\,, \qquad C_0(s,t) = st + (st)^{1/2}\,,$$

$$c_1 = \int_0^1 (s^2 + s)\, ds = \tfrac{5}{6}\,,$$

$$C_1(s,t) = \tfrac{5}{6}[st + (st)^{1/2}] - \int_0^1 [sx + (sx)^{1/2}][xt + (xt)^{1/2}]\, dt$$

$$= \tfrac{1}{2}st + \tfrac{1}{3}(st)^{1/2} - \tfrac{2}{5}(st^{1/2} + ts^{1/2})\,,$$

$$c_2 = \int_0^1 (\tfrac{1}{2}s^2 + \tfrac{1}{3}s - \tfrac{4}{5}s^{3/2})\, ds = 1/75\,,$$

$$C_2(s,t) = 0\,,$$

and therefore all the subsequent coefficients vanish. The value of the resolvent is

$$\Gamma(s,t;\lambda) = \frac{st + (st)^{1/2} - \{\tfrac{1}{2}st + \tfrac{1}{3}(st)^{1/2} - \tfrac{2}{5}(st^{1/2} + s^{1/2}t)\}\lambda}{1 - \tfrac{5}{6}\lambda + (1/150)\lambda^2}\;. \tag{8}$$

The solution $g(s)$ then follows by using the relation (4.2.23),

$$g(s) = \frac{150s + \lambda(60\sqrt{s} - 75s) + 21\lambda^2 s}{\lambda^2 - 125\lambda + 150}\;. \tag{9}$$

Example 3.   As a final example, let us solve the integral equation

$$g(s) = 1 + \lambda \int_0^\pi [\sin(s+t)] g(t) \, dt \,, \tag{10}$$

which we have already solved in Section 3.2 by a Neumann series. For the present kernel,

$$c_0 = 1, \qquad C_0(s,t) = \sin(s+t), \qquad c_1 = \int_0^\pi \sin 2s \, ds = 0 \,.$$

$$C_1(s,t) = 0 - \int_0^\pi \sin(s+x)\sin(x+t) \, dx = -\tfrac{1}{2}\pi \cos(s-t) \,,$$

$$c_2 = -\int_0^\pi \tfrac{1}{2}\pi \cos 0 \, ds = -\tfrac{1}{2}\pi^2 \,,$$

$$C_2(s,t) = -\tfrac{1}{2}\pi^2 \sin(s+t) + 2 \int_0^\pi \tfrac{1}{2}\pi \sin(s+x)\cos(x-t) \, dx = 0 \,.$$

Hence, the resolvent is

$$\Gamma(s,t;\lambda) = \frac{\sin(s+t) + \tfrac{1}{2}\pi\lambda \cos(s-t)}{1 - \tfrac{1}{4}\pi^2 \lambda^2} \,,$$

and the solution is

$$g(s) = 1 + \frac{2\lambda(\cos s) + \lambda^2 \pi(\sin s)}{1 - \tfrac{1}{4}\pi^2 \lambda^2} \,, \tag{11}$$

which agrees with the solution (3.2.10).

4.4. FREDHOLM'S SECOND THEOREM

Fredholm's first theorem does not hold when λ is a root of the equation $D(\lambda) = 0$. We have found in Chapter 2 that, for a separable kernel, the homogeneous equation

$$g(s) = \lambda \int K(s,t) g(t) \, dt \tag{1}$$

has nontrivial solutions. It might be expected that same holds when the kernel is an arbitrary integrable function and we shall then have a spectrum of eigenvalues and corresponding eigenfunctions. The second theorem of Fredholm is devoted to the study of this problem.

We first prove that every zero of $D(\lambda)$ is a pole of the resolvent kernel (4.2.24); the order of this pole is at most equal to the order of the zero of $D(\lambda)$. In fact, differentiate the Fredholm's first series (4.2.26) and interchange the indices of the variables of integration to get

$$D'(\lambda) = - \int D(s,s;\lambda)\, ds \ . \tag{2}$$

From this relation, it follows that, if λ_0 is a zero of order k of $D(\lambda)$, then it is a zero of order $k-1$ of $D'(\lambda)$ and consequently λ_0 may be a zero of order at most $k-1$ of the entire function $D(s,t;\lambda)$. Thus, λ_0 is the pole of the quotient (4.2.24) of order at most k. In particular, if λ_0 is a simple zero of $D(\lambda)$, then $D(\lambda_0) = 0$, $D'(\lambda_0) \neq 0$, and λ_0 is a simple pole of the resolvent kernel. Moreover, it follows from (2) that $D(s,t;\lambda) \neq 0$. For this particular case, we observe from equation (4.1.8) that, if $D(\lambda) = 0$ and $D(s,t;\lambda) \neq 0$, then $D(s,t;\lambda)$, as a function of s, is a solution of the homogeneous equation (1). So is $\alpha D(s,t;\lambda)$, where α is an arbitrary constant.

Let us now consider the general case when λ is a zero of an arbitrary multiplicity m, that is, when

$$D(\lambda_0) = 0\,, \qquad \ldots\,, \qquad D^{(r)}(\lambda_0) = 0\,, \qquad D^{(m)}(\lambda_0) \neq 0\,, \tag{3}$$

where the superscript r stands for the differential of order r, $r = 1$, $\ldots$, $m-1$. For this case, the analysis is simplified if one defines a determinant known as the Fredholm minor:

$$\begin{aligned}
D_n &\begin{pmatrix} s_1, s_2, \ldots, s_n \\ t_1, t_2, \ldots, t_n \end{pmatrix} \lambda \\
&= K\begin{pmatrix} s_1, s_2, \ldots, s_n \\ t_1, t_2, \ldots, t_n \end{pmatrix} + \sum_{p=1}^{\infty} \frac{(-\lambda)^p}{p!} \\
&\quad \times \int \cdots \int K\begin{pmatrix} s_1, \ldots, s_n, x_1, \ldots, x_p \\ t_1, \ldots, t_n, x_1, \ldots, x_p \end{pmatrix} dx_1\, dx_2 \cdots dx_p\,,
\end{aligned} \tag{4}$$

where $\{s_i\}$ and $\{t_i\}$, $i = 1, 2, \ldots, n$, are two sequences of arbitrary variables. Just as do the Fredholm series (4.2.25) and (4.2.26), the series (4) also converges for all values of λ and consequently is an entire function of λ. Furthermore, by differentiating the series (4.2.26) n times and comparing it with the series (4), there follows the relation

$$\frac{d^n D(\lambda)}{d\lambda^n} = (-1)^n \int \cdots \int D_n \left(\begin{matrix} s_1, \ldots, s_n \\ s_1, \ldots, s_n \end{matrix} \bigg| \lambda \right) ds_1 \cdots ds_n \ . \tag{5}$$

From this relation, we conclude that, if λ_0 is a zero of multiplicity m of the function $D(\lambda)$, then the following holds for the Fredholm minor of order m for that value of λ_0:

$$D_m \left(\begin{matrix} s_1, s_2, \ldots, s_m \\ t_1, t_2, \ldots, t_m \end{matrix} \bigg| \lambda_0 \right) \neq 0 \ .$$

Of course, there might exist minors of order lower than m which also do not identically vanish (compare the discussion in Section 2.3).

Let us find the relation among the minors that corresponds to the resolvent formula (4.2.7). Expansion of the determinant under the integral sign in (4),

$$\begin{vmatrix} K(s_1, t_1) & K(s_1, t_2) & \cdots & K(s_1, t_n) & K(s_1, x_1) & \cdots & K(s_1, x_p) \\ K(s_2, t_1) & K(s_2, t_2) & \cdots & K(s_2, t_n) & K(s_2, x_1) & \cdots & K(s_2, x_p) \\ \vdots & & & & & & \\ K(s_n, t_1) & K(s_n, t_2) & \cdots & K(s_n, t_n) & K(s_n, x_1) & \cdots & K(s_n, x_p) \\ K(x_1, t_1) & K(x_1, t_2) & \cdots & K(x_1, t_n) & K(x_1, x_1) & \cdots & K(x_1, x_p) \\ \vdots & & & & & & \\ K(x_p, t_1) & K(x_p, t_2) & \cdots & K(x_p, t_n) & K(x_p, x_1) & \cdots & K(x_p, x_p) \end{vmatrix} \tag{6}$$

by elements of the first row and integrating p times with respect to $x_1, x_2, \ldots, x_p$ for $p \geqslant 1$, we have

$$\int \cdots \int K \left(\begin{matrix} s_1, \ldots, s_n, x_1, \ldots, x_p \\ t_1, \ldots, t_n, x_1, \ldots, x_p \end{matrix} \right) dx_1 \cdots dx_p$$

$$= \sum_{h=1}^{n} (-1)^{h+1} K(s_1, t_h)$$

$$\times \int \cdots \int K \left(\begin{matrix} s_2, \ldots, s_h, \ldots, & s_n, x_1, \ldots, x_p \\ t_1, \ldots, t_{h-1}, t_{h+1}, \ldots, t_n, x_1, \ldots, x_p \end{matrix} \right) dx_1 \, dx_2 \cdots dx_p$$

$$+ \sum_{h=1}^{p} (-1)^{h+n-1}$$

$$\times \int \cdots \int K(s_1, x_h) K \left(\begin{matrix} s_2, \ldots, & s_n, x_1, x_2, \ldots, x_h, \ldots, x_p \\ t_1, \ldots, t_{n-1}, t_n, x_1, \ldots, x_{h-1}, x_{h+1}, \ldots, x_p \end{matrix} \right)$$

$$\times dx_1 \cdots dx_p \ . \tag{7}$$

Note that the symbols for the determinant K on the right side of (7) do not contain the variables s_1 in the upper sequence and the variables t_h or x_h in the lower sequence. Furthermore, it follows by transposing the variable s_h in the upper sequence to the first place by means of $h+n-2$ transpositions that all the components of the second sum on the right side are equal. Therefore, we can write (7) as

$$
\int \cdots \int K\begin{pmatrix} s_1, \ldots, s_n, x_1, \ldots, x_p \\ t_1, \ldots, t_n, x_1, \ldots, x_p \end{pmatrix}
$$

$$
= \sum_{h=1}^{n} (-1)^{h+1} K(s_1, t_h)
$$

$$
\times \int \cdots \int K\begin{pmatrix} s_2, \ldots, & s_n, x_1, \ldots, x_p \\ t_1, \ldots, t_{h-1}, t_{h+1}, \ldots, t_n, x_1, \ldots, x_p \end{pmatrix} dx_1 \cdots dx_p
$$

$$
- p \int K(s_1, x)\left[\int \cdots \int K\begin{pmatrix} x, s_2, \ldots, s_n, x_1, \ldots, x_{p-1} \\ t_1, t_2, \ldots, t_n, x_1, \ldots, x_{p-1} \end{pmatrix}\right.
$$

$$
\left. \times dx_1 \cdots dx_{p-1} \right] dx \tag{8}
$$

where we have omitted the subscript h from x. Substituting (8) in (7), we find that Fredholm minor satisfies the integral equation

$$
D_n\begin{pmatrix} s_1, \ldots, s_n \\ t_1, \ldots, t_n \end{pmatrix} \lambda) = \sum_{h=1}^{n} (-1)^{h+1} K(s_1, t_h) D_{n-1}\begin{pmatrix} s_2, \ldots & , s_n \\ t_1, \ldots, t_{h-1}, t_{h+1}, t_n \end{pmatrix}
$$

$$
+ \lambda \int K(s_1, x) D_n\begin{pmatrix} x, x_2, \ldots, s_n \\ t_1, t_2, \ldots, t_n \end{pmatrix} \lambda) dx . \tag{9}
$$

Expansion by the elements of any other row leads to a similar identity, with x placed at the corresponding place. If we expand the determinant (6) with respect to the first column and proceed as above, we get the integral equation

$$
D_n\begin{pmatrix} s_1, \ldots, s_n \\ t_1, \ldots, t_n \end{pmatrix} \lambda) = \sum_{h=1}^{n} (-1)^{h+1} K(s_h, t_1) D_{n-1}\begin{pmatrix} s_1, \ldots, s_{h-1}, s_{h+1}, \ldots, s_n \\ t_2, \ldots & , t_n \end{pmatrix}
$$

$$
+ \lambda \int K(x, t_1) D_n\begin{pmatrix} s_1, \ldots & , s_n \\ x, t_2, \ldots, t_n \end{pmatrix} dx , \tag{10}
$$

and a similar result would follow if we were to expand by any other column. The formulas (9) and (10) will play the role of the Fredholm series of the previous section.

Note that the relations (9) and (10) hold for all values of λ. With the help of (9), we can find the solution of the homogeneous equation (1) for the special case when $\lambda = \lambda_0$ is an eigenvalue. To this end, let us suppose that $\lambda = \lambda_0$ is a zero of multiplicity m of the function $D(\lambda)$. Then, as remarked earlier, the minor D_m does not identically vanish and even the minors $D_1, D_2, \ldots, D_{m-1}$ may not identically vanish. Let D_r be the first minor in the sequence $D_1, D_2, \ldots, D_{m-1}$ that does not vanish identically. The number r lies between 1 and m and is the index of the eigenvalue λ_0 as defined in Section 2.3. Moreover, this means that $D_{r-1} = 0$. But then the integral equation (9) implies that

$$g_1(s) = D_r \begin{pmatrix} s, s_2, \ldots, s_r \\ t_1, \ldots, \quad t_r \end{pmatrix} \lambda_0 \tag{11}$$

is a solution of the homogeneous equation (1). Substituting s at different points of the upper sequence in the minor D_r, we obtain r nontrivial solutions $g_i(s)$, $i = 1, \ldots, r$, of the homogeneous equation. These solutions are usually written as

$$\Phi_i(s) = \frac{D_r \begin{pmatrix} s_1, \ldots, s_{i-1}, s, s_{i+1}, \ldots, s_r \\ t_1, \ldots \qquad\qquad\qquad , t_r \end{pmatrix} \lambda_0}{D_r \begin{pmatrix} s_1, \ldots, s_{i-1}, s_i, s_{i+1}, \ldots, s_r \\ t_1, \ldots \qquad\qquad\qquad , t_r \end{pmatrix} \lambda_0}, \qquad i = 1, 2, \ldots, r. \tag{12}$$

Observe that we have already established that the denominator is not zero.

The solutions Φ_i as given by (12) are linearly independent for the following reason. In the determinant (6) above, if we put two of the arguments s_i equal, this amounts to putting two rows equal, and consequently the determinant vanishes. Thus, in (12), we see that $\Phi_k(s_i) = 0$ for $i \neq k$, whereas $\Phi_k(s_k) = 1$. Now, if there exists a relation $\sum_k C_k \Phi_k \equiv 0$, we may put $s = s_i$, and it follows that $C_i \equiv 0$; and this proves the linear independence of these solutions. This system of solutions Φ_i is called the fundamental system of the eigenfunctions of λ_0 and any linear combination of these functions gives a solution of (1).

Conversely, we can show that any solution of equation (1) must be a linear combination of $\Phi_1(s), \Phi_2(s), \ldots, \Phi_\nu(s)$. We need to define a kernel $H(s, t; \lambda)$ which corresponds to the resolvent kernel $\Gamma(s, t; \lambda)$ of the previous section

$$H(s, t; \lambda) = D_{r+1}\left(\begin{matrix} s, s_1, \ldots, s_r \\ t, t_1, \ldots, t_r \end{matrix}\middle| \lambda_0\right) \middle/ D_r\left(\begin{matrix} s_1, \ldots, s_r \\ t_1, \ldots, t_r \end{matrix}\middle| \lambda_0\right). \tag{13}$$

In (10), take n to be equal to r, and add extra arguments s and t to obtain

$$D_{r+1}\left(\begin{matrix} s, s_1, \ldots, s_r \\ t, t_1, \ldots, t_r \end{matrix}\middle| \lambda_0\right) = K(s, t) D_r\left(\begin{matrix} s_1, \ldots, s_r \\ t_1, \ldots, t_r \end{matrix}\middle| \lambda_0\right)$$

$$+ \sum_{h=1}^{r} (-1)^h K(s_h, y) D_r\left(\begin{matrix} s, s_1, \ldots, s_{h-1}, s_{h+1}, \ldots, s_r \\ t_1, t_2, \ldots \qquad\qquad , t_r \end{matrix}\middle| \lambda_0\right)$$

$$+ \lambda_0 \int K(x, t) D_{r+1}\left(\begin{matrix} s, s_1, \ldots, s_r \\ x, t_1, \ldots, t_r \end{matrix}\middle| \lambda_0\right) dx . \tag{14}$$

In every minor D_r in the above equation, we transpose the variable s from the first place to the place between the variables s_{h-1} and s_{h+1} and divide both sides by the constant

$$D_r\left(\begin{matrix} s_1, \ldots, s_r \\ t_1, \ldots, t_r \end{matrix}\middle| \lambda_0\right) \neq 0 ,$$

to obtain

$$H(s, t; \lambda) - K(s, t) - \lambda_0 \int H(s, x; \lambda) K(x, t) \, dx$$

$$= - \sum_{h=1}^{r} K(s_h, t) \Phi_h(s) . \tag{15}$$

If $g(s)$ is any solution to (1), we multiply (15) by $g(t)$ and integrate with respect to t,

$$\int g(t) H(s, t; \lambda) \, dt - \frac{g(s)}{\lambda_0} - \int g(x) \Gamma(s, x; \lambda) \, dx$$

$$= - \sum_{h=1}^{r} \frac{g(s_h)}{\lambda_0} \Phi_h(s) , \tag{16}$$

where we have used (1) in all terms but the first; we have also taken $\lambda_0 \int K(s_h, t) g(t) \, dt = g(s_h)$. Cancelling the equal terms, we have

$$g(s) = \sum_{h=1}^{r} g(s_h) \Phi_h(s) . \tag{17}$$

This proves our assertion. Thus we have established the following result.

Fredholm's Second Theorem. If λ_0 is a zero of multiplicity m of the function $D(\lambda)$, then the homogeneous equation

$$g(s) = \lambda_0 \int K(s, t) g(t) \, dt \tag{18}$$

possesses at least one, and at most m, linearly independent solutions

$$g_i(s) = D_r \left(\begin{matrix} s_1, \ldots, s_{i-1}, s, s_{i+1}, \ldots, s_r \\ t_1, \ldots \end{matrix} \middle| \begin{matrix} \lambda_0 \\ , t_r \end{matrix} \right),$$

$$i = 1, \ldots, r ; \quad 1 \leqslant r \leqslant m \tag{19}$$

not identically zero. Any other solution of this equation is a linear combination of these solutions.

4.5. FREDHOLM'S THIRD THEOREM

In the analysis of Fredholm's first theorem, it has been shown that the inhomogeneous equation

$$g(s) = f(s) + \lambda \int K(s, t) g(t) \, dt \tag{1}$$

possesses a unique solution provided $D(\lambda) \neq 0$. Fredholm's second theorem is concerned with the study of the homogeneous equation

$$g(s) = \lambda \int K(s, t) g(t) \, dt ,$$

when $D(\lambda) = 0$. In this section, we investigate the possibility of (1) having a solution when $D(\lambda) = 0$. The analysis of this section is not much different from the corresponding analysis for separable kernels as given in Section 2.3. In fact, the only difference is that we shall now give an explicit formula for the solution. Qualitatively, the discussion is the same.

Recall that the transpose (or adjoint) of equation (1) is (under the same assumption as in Section 2.3)

$$\psi(s) = f(s) + \lambda \int_a^b K(t,s)\psi(t)\,dt\ . \tag{2}$$

It is clear that Fredholm's first series $D(\lambda)$ as given by (4.1.26) is the same for the transposed equation, while the second series is $D(t,s;\lambda)$ as obtained from (4.1.25) by interchanging the roles of s and t. This means that the kernels of equation (1) and its transpose (2) have the same eigenvalues. Furthermore, the resolvent kernel for (2) is

$$\Gamma(t,s;\lambda) = D(t,s;\lambda)/D(\lambda)\ , \tag{3}$$

and therefore the solution of (2) is

$$\psi(s) = f(s) + \lambda \int [D(t,s;\lambda)/D(\lambda)]f(t)\,dt\ , \tag{4}$$

provided λ is not an eigenvalue.

It is also clear that not only has the transposed kernel the same eigenvalues as the original kernel, but also the index r of each of the eigenvalues is equal. Moreover, corresponding to equation (4.4.12), the eigenfunctions of the transposed equation for an eigenvalue for λ_0 are given as

$$\Psi_i(t) = \frac{D_r\left(\begin{matrix} s_1,\ \ldots\ \ \ \ \ \ \ \ \ \ \ \ \ \ \ \ \ ,s_r \\ t_1,\ \ldots,\ t_{i-1},\ t,\ t_{i+1},\ \ldots,\ t_r \end{matrix}\ \middle|\ \lambda_0\right)}{D_r\left(\begin{matrix} s_1,\ \ldots\ \ \ \ \ \ \ \ \ \ \ \ \ \ \ \ \ ,s_r \\ t_1,\ \ldots,\ t_{i-1},\ t_i,\ t_{i+1},\ \ldots,\ t_r \end{matrix}\ \middle|\ \lambda_0\right)}\ , \tag{5}$$

where the values $(s_1,\ldots,s_r)$ and $(t_1,\ldots,t_r)$ are so chosen that the denominator does not vanish. Substituting r in different places in the lower sequence of this formula, we obtain a linearly independent system of r eigenfunctions. Also recall that each Φ_i is orthogonal to each Ψ_i with different eigenvalues.

If a solution $g(s)$ of (1) exists, then multiply (1) by each member $\Psi_k(s)$ of the above-mentioned system of functions and integrate to obtain

$$\int f(s)\,\Psi_k(s)\,ds = \begin{aligned} &\int g(s)\,\Psi_k(s)\,ds - \lambda \iint K(s,t)\,g(t)\,\Psi_k(s)\,ds\,dt \\ &\int g(s)\,ds\,[\Psi_k(s) - \lambda \int K(t,s)\,\Psi_k(t)\,dt] = 0\ , \end{aligned} \tag{6}$$

where the term in the bracket vanishes because $\Psi_k(s)$ is an eigenfunction of the transposed equation. From (6), we see that a necessary condition for (1) to have a solution is that the inhomogeneous term $f(s)$ be orthogonal to each solution of the transposed homogeneous equation.

Conversely, we shall show that the condition (6) of orthogonality is sufficient for the existence of a solution. Indeed, we shall present an explicit solution in that case. With this purpose, we again appeal to the resolvent function $H(s, t; \lambda)$ as defined by (4.4.13) under the assumption that $D_r \neq 0$ and that r is the index of the eigenvalue λ_0.

Our contention is that if the orthogonality condition is satisfied, then the function

$$g_0(s) = f(s) + \lambda_0 \int H(s, t; \lambda) f(t) \, dt \tag{7}$$

is a solution. Indeed, substitute this value for $g(s)$ in (1), obtaining

$$f(s) + \lambda_0 \int H(s, t; \lambda) f(t) \, dt = f(s) + \lambda_0 \int K(s, t)$$
$$\times [f(t) + \lambda_0 \int H(t, x; \lambda) f(x) \, dx] \, dt$$

or

$$\int f(t) \, dt \, [H(s, t; \lambda) - K(s, t) - \lambda_0 \int K(s, x) H(x, t; \lambda) \, dx] = 0 . \tag{8}$$

Now, just as we obtained equation (4.4.15), we can obtain its "transpose,"

$$H(s, t; \lambda) - K(s, t) - \lambda_0 \int K(s, x) H(x, t; \lambda) \, dx$$
$$= -\sum_{h=1}^{r} K(s, t_h) \Psi_h(t) . \tag{9}$$

Substituting this in (8) and using the orthogonality condition, we have an identity, and thereby the assertion is proved.

The difference of any two solutions of (1) is a solution of the homogeneous equation. Hence, the most general solution of (1) is

$$g(s) = f(s) + \lambda_0 \int H(s, t; \lambda) f(t) \, dt + \sum_{h=1}^{r} C_h \Phi_h(s) . \tag{10}$$

The above analysis leads to the following theorem.

Fredholm's Third Theorem. For an inhomogeneous equation

$$g(s) = f(s) + \lambda_0 \int K(s, t) g(t) \, dt , \tag{11}$$

to possess a solution in the case $D(\lambda_0) = 0$, it is necessary and sufficient that the given function $f(s)$ be orthogonal to all the eigenfunctions

$\Psi_i(s)$, $i = 1, 2, ..., \nu$, of the transposed homogeneous equation corresponding to the eigenvalue λ_0. The general solution has the form

$$g(s) = f(s) + \lambda_0 \int \left[D_{r+1} \begin{pmatrix} s, s_1, s_2, ..., s_r \\ t, t_1, ... \quad , t_r \end{pmatrix} \lambda_0 \right] \bigg/ D_r \begin{pmatrix} s_1, s_2, ..., s_r \\ t_1, ... \quad , t_r \end{pmatrix} \lambda_0 \right)$$

$$\times f(t) \, dt + \sum_{h=1}^{r} C_h \Phi_h(s) . \tag{12}$$

EXERCISES

1. Use the method of this chapter and find the resolvent kernels for the following three integral equations:

$$\text{(i)} \quad g(s) = f(s) + \lambda \int_0^1 |s-t| g(t) \, dt ,$$

$$\text{(ii)} \quad g(s) = f(s) + \lambda \int_0^1 (\exp - |s-t|) g(t) \, dt ,$$

$$\text{(iii)} \quad g(s) = f(s) + \lambda \int_0^{2\pi} [\cos(s+t)] g(t) \, dt .$$

2. Solve the following homogeneous equations

$$\text{(i)} \quad g(s) = \tfrac{1}{2} \int_0^{\pi} [\sin(s+t)] g(t) \, dt ,$$

$$\text{(ii)} \quad g(s) = [1/(e^2 - 1)] \int_0^1 2e^s e^t g(t) \, dt ,$$

by using the explicit formulas of this chapter.

3. Show by the present method that the resolvent kernel for the integral equation with kernel $K(s, t) = 1 - 3st$, in the interval $(0, 1)$, is

$$\Gamma(s, t; \lambda) = [4/(4 - \lambda^2)] [1 + \lambda - \tfrac{3}{2}(s+t) - 3(1-\lambda) st] , \qquad \lambda \neq \pm 2 .$$

4. Show that not every one of the Fredholm's minors as defined by (4.4.4) is identically zero.

Hint: Use (4.4.5).

APPLICATIONS TO ORDINARY DIFFERENTIAL EQUATIONS

The theories of ordinary and partial differential equations are fruitful sources of integral equations. In the quest for the representation formula for the solution of a linear differential equation in such a manner so as to include the boundary condition or initial condition explicitly, one is always led to an integral equation. Once a boundary value or an initial value problem has been formulated in terms of an integral equation, it becomes possible to solve this problem easily. In this chapter, we shall consider only ordinary differential equations. The next chapter is devoted to partial differential equations.

5.1. INITIAL VALUE PROBLEMS

There is a fundamental relationship between Volterra integral equations and ordinary differential equations with prescribed initial values. We begin our discussion by studying the simple initial value problem

$$y'' + A(s)y' + B(s)y = F(s) , \tag{1}$$

$$y(a) = q_0 , \qquad y'(a) = q_1 , \tag{2}$$

where a prime implies differentiation with respect to s, and the functions A, B, and F are defined and continuous in the closed interval $a \leqslant s \leqslant b$.

The result of integrating the differential equation (1) from a to s and using the initial values (2) is

$$y'(s) - q_1 = - A(s) y(s) - \int_a^s [B(s_1) - A'(s_1)] y(s_1) \, ds_1$$

$$+ \int_a^s F(s_1) \, ds_1 + A(a) q_0 .$$

Similarly, a second integration yields

$$y(s) - q_0 = - \int_a^s A(s_1) y(s_1) \, ds_1 - \int_a^s \int_a^{s_2} [B(s_1) - A'(s_1)] y(s_1) \, ds_1 \, ds_2$$

$$+ \int_a^s \int_a^{s_2} F(s_1) \, ds_1 \, ds_2 + [A(a) q_0 + q_1] (s - a) . \tag{3}$$

With the help of the identity (see Appendix, Section A.1)

$$\int_a^s \int_a^{s_2} F(s_1) \, ds_1 \, ds_2 = \int_a^s (s - t) F(t) \, dt , \tag{4}$$

the two double integrals in (3) can be converted to single integrals. Hence, the relation (3) takes the form

$$y(s) = q_0 + [A(a) q_0 + q_1] (s - a) + \int_a^s (s - t) F(t) \, dt$$

$$- \int_a^s \{A(t) + (s - t) [B(t) - A'(t)]\} y(t) \, dt . \tag{5}$$

Now, set

$$K(s, t) = - \{A(t) + (s - t) [B(t) - A'(t)]\} \tag{6}$$

and

$$f(s) = \int_a^s (s - t) F(t) \, dt + [A(a) q_0 + q_1] (s - a) + q_0 . \tag{7}$$

From relations (5)–(7), we have the Volterra integral equation of the second kind:

$$y(s) = f(s) + \int_a^s K(s, t) y(t) \, dt \, . \tag{8}$$

Conversely, any solution $g(s)$ of the integral equation (8) is, as can be verified by two differentiations, a solution of the initial value problem (1)–(2).

Note that the crucial step is the use of the identity (4). Since we have proved the corresponding identity for an arbitrary integer n in the Appendix, Section A.1, it follows that the above process of converting an initial value problem to a Volterra integral equation is applicable to a linear ordinary differential equation of order n when there are n prescribed initial conditions. An alternative approach is somewhat simpler for proving the above-mentioned equivalence for a general differential equation. Indeed, let us consider the linear differential equation of order n:

$$\frac{d^n y}{ds^n} + A_1(s)\frac{d^{n-1} y}{ds^{n-1}} + \cdots + A_{n-1}(s)\frac{dy}{ds} + A_n(s)y = F(s) \, , \tag{9}$$

with the initial conditions

$$y(a) = q_0 \, , \qquad y'(a) = q_1, \qquad \ldots, \qquad y^{(n-1)}(a) = q_{n-1} \, , \tag{10}$$

where the functions $A_1, A_2, \ldots, A_n$ and F are defined and continuous in $a \leqslant s \leqslant b$.

The reduction of the initial value problem (9)–(10) to the Volterra integral equation is accomplished by introducing an unknown function $g(s)$:

$$d^n y/ds^n = g(s) \, . \tag{11}$$

From (10) and (11), it follows that

$$\frac{d^{n-1}(y)}{ds^{n-1}} = \int_a^s g(t) \, dt + q_{n-1} \, ,$$

$$\frac{d^{n-2} y}{ds^{n-2}} = \int_a^s (s-t)g(t) \, dt + (s-a)q_{n-1} + q_{n-2} \, ,$$

$$\tag{12}$$

(continued)

$$y(0) = 0 = y(1), \qquad y''(0) = 0 = y'''(1)$$

to a Fredholm integral equation.

12. Extend the theory of Section 5.8 to the case when the completely homogeneous system has two linearly independent solutions $w_1(s)$ and $w_2(s)$.

13. Find the modified Green's function for the systems

$$y'' - \lambda y = 0, \qquad y(0) = y(1), \qquad y'(0) = y'(1)$$

14. Transform the boundary value problem

$$y'' + y = f(s), \qquad y(0) = y(\pi) = 0,$$

into an integral equation.

Hint: The self-adjoint operator $-(y''+y) = 0$ has the nontrivial solution $(2/\pi)^{\frac{1}{2}} \sin s$ and it satisfies the boundary conditions. Now proceed as in the examples in Section 5.9.

APPLICATIONS TO PARTIAL DIFFERENTIAL EQUATIONS

CHAPTER 6

6.1. INTRODUCTION

The applications of integral equations are not restricted to ordinary differential equations. In fact, the most important applications of integral equations arise in finding the solutions of boundary value problems in the theory of partial differential equations of the second order. The boundary value problems for equations of elliptic type can be reduced to Fredholm integral equations, while the study of parabolic and hyperbolic differential equations leads to Volterra integral equations. We shall confine our attention to the linear partial differential equations of the elliptic type, specifically, to the Laplace, Poisson, and Helmholtz equations, wherein lie the most interesting and important achievements of the theory of integral equations.

Three types of boundary conditions arise in the study of elliptic partial differential equations. The first type is the Dirichlet condition. In this case, we prescribe the value of the solution on the boundary. The second type is the Neumann condition. In this case, we prescribe the normal derivative of the solution on the boundary. When we prescribe the Dirichlet conditions on some parts and the Neumann conditions on other parts of the boundary, we have a mixed boundary value problem.

94

The divergence theorem and the two Green's identities will be used repeatedly in this chapter. They are as follows:

Divergence theorem:

$$\int_R \text{div } \mathbf{A} \, dV = \int_S \mathbf{A} \cdot \mathbf{n} \, dS \; ; \tag{1}$$

Green's first identity:

$$\int_R u\nabla^2 v \, dV = -\int_R (\text{grad } u \cdot \text{grad } v) \, dV + \int_S u \frac{\partial v}{\partial n} \, dS \; ; \tag{2}$$

Green's second identity:

$$\int_R (u\nabla^2 v - v\nabla^2 u) \, dV = \int_S \left(u \frac{\partial v}{\partial n} - v \frac{\partial u}{\partial n} \right) dS \; , \tag{3}$$

where $\mathbf{A}$ is a continuously differentiable vector field and the functions u and v have partial derivatives of the second order which are continuous in the bounded region R; S is the boundary of R, while $\mathbf{n}$ stands for the unit normal outward to S. The surface S is a smooth (or regular) surface as defined by Kellog [9]. The differential operator ∇^2 is the Laplacian, which in Cartesian coordinates x_1, x_2, x_3 has the form

$$\nabla^2 = \frac{\partial^2}{\partial x_1^2} + \frac{\partial^2}{\partial x_2^2} + \frac{\partial^2}{\partial x_3^2} \; . \tag{4}$$

Any solution u of the equation $\nabla^2 u = 0$ is called a harmonic function.

To keep the chapter to a manageable size, we shall confine ourselves mainly to three-dimensional problems. The results can be readily extended to two-dimensional problems once the technique is grasped. Boldface letters such as $\mathbf{x}$ shall signify the triplet (x_1, x_2, x_3). The quantities R_i and R_e will stand for the regions interior and exterior to S, respectively. Furthermore, we shall not follow the common practice of writing dS_x or dS_ξ to signify that integration is with respect to the variable $\mathbf{x}$ or ξ. We shall merely write dS and it shall be clear from the context as to what the variable of integration is.

We shall be interested in giving the integral-equation formulation mainly to the differential equations $\nabla^2 u = 0$, the Laplace equation;

$\nabla^2 u = -4\pi\rho$, the Poisson equation; and $(\nabla^2 + k^2)u = 0$, the Helmholtz (or the steady-state wave) equation. Here, $\rho(\mathbf{x})$ is a given function of position and k is a given number.

6.2. INTEGRAL REPRESENTATION FORMULAS FOR THE SOLUTIONS OF THE LAPLACE AND POISSON EQUATIONS

Our starting point is the fundamental solution (or free-space solution) $E(\mathbf{x}; \xi)$ which satisfies

$$-\nabla^2 E = \delta(\mathbf{x} - \xi) \tag{1}$$

and vanishes at infinity. This function can be interpreted as the electrostatic potential at an arbitrary field point $\mathbf{x}$ due to a unit charge at the source point ξ. Such a potential is given as

$$E(\mathbf{x}; \xi) = 1/4\pi r = 1/4\pi |\mathbf{x} - \xi| . \tag{2}$$

For the two-dimensional case, the corresponding formula is $(1/2\pi)\log(1/r) = (1/2\pi)\log(1/|\mathbf{x} - \xi)$, where $\mathbf{x} = (x_1, x_2)$ and $\xi = (\xi_1, \xi_2)$.

The fundamental solution can be used to obtain the solution of the Poisson equation

$$-\nabla^2 u = 4\pi\rho . \tag{3}$$

Indeed, multiply (1) by $u(x)$, (3) by $E(\mathbf{x}; \xi)$, subtract, integrate over the region R_i, and use Green's second identity as well as the sifting property of the delta function. The result is (after relabeling $\mathbf{x}$ and ξ)

$$u(\mathbf{x}) = \int_R \frac{\rho}{r} \, dV - \frac{1}{4\pi} \int_S u \frac{\partial}{\partial n}\left(\frac{1}{r}\right) dS + \frac{1}{4\pi} \int_S \frac{1}{r} \frac{\partial u}{\partial n} \, dS . \tag{4}$$

Suppose that from some previous considerations we know the values of ρ, u, and $\partial u/\partial n$ which appear in the formula (4):

$$u|_S = \tau , \qquad \partial u/\partial n|_S = \sigma ; \tag{5}$$

then this formula becomes

$$u(P) = \int_R \frac{\rho}{r} \, dV + \frac{1}{4\pi} \int_S \tau(Q) \frac{\partial}{\partial n}\left(\frac{1}{r}\right) dS - \frac{1}{4\pi} \int_S \frac{\sigma(Q)}{r} \, dS , \tag{6}$$

where P is the field point $\mathbf{x}$ and Q is a point ξ on S. The formulas (4) and (6) lead to many interesting properties of the harmonic functions. For details, the reader should refer to the appropriate texts [4, 9, 19]. We shall merely note the properties of the three integrals which occur on the right side of (6).

The Newtonian, Single-Layer, and Double-Layer Potentials

The integral $\int_R (\rho/r)\, dV$ is called the volume potential or Newtonian potential with volume density ρ. Similarly, the integral $\int_S (\sigma/r)\, dS$ is the simple or single-layer potential with charge (or source) density σ, while the integral $\int_S \tau(\partial/\partial n)(1/r)\, dS$ is the double-layer potential with dipole density τ. These integrals arise in all the fields of potential theory. However, we have used the language of electrostatics, as it interprets them nicely. These potentials have the following properties.

For the Newtonian potential $u = \int_R (\rho/r)\, dV$, we have the following:

(1) $\nabla^2 u = 0$, for points P in R_e.

(2) For points P within R_i, the integral is improper but it converges and admits two differentials under the integral sign if the function ρ is sufficiently smooth; the result is $\nabla^2 u = -4\pi\rho(P)$.

The single-layer potential $u = \int_S (\sigma/r)\, dS$ has the following properties:

(1) $\nabla^2 u = 0$, outside S.

(2) The integral becomes improper at the surface S but converges uniformly if S is regular. Moreover, this integral remains continuous as we pass through S.

(3) Consider the derivative of u taken in the direction of a line normal to the surface S in the outward direction from S. Then,

$$\frac{\partial u}{\partial n}\bigg|_{P_+} = -2\pi\sigma(P) + \int_S \sigma(Q)\,\frac{\cos(\xi-\mathbf{x},\mathbf{n})}{|\mathbf{x}-\xi|^2}\, dS \tag{7}$$

and

$$\frac{\partial u}{\partial n}\bigg|_{P_-} = 2\pi\sigma(P) + \int_S \sigma(Q)\,\frac{\cos(\xi-\mathbf{x},\mathbf{n})}{|\mathbf{x}-\xi|^2}\, dS\,, \tag{8}$$

where P_+ and P_- signify that we approach S from R_i and R_e, respectively, and where both $\mathbf{x}$ and ξ are on S. From (7) and (8), we obtain the jump of the normal derivative of u across S:

$$\sigma = \frac{1}{4\pi}\left(\frac{\partial u}{\partial n}\bigg|_{P_-} - \frac{\partial u}{\partial n}\bigg|_{P_+}\right). \tag{9}$$

Similarly, the double-layer potential $u = \int_S \tau(\partial/\partial n)(1/r)\,dS$, has the following properties:

(1) $\nabla^2 u = 0$ outside of S.

(2)   The integral becomes improper at the surface but it converges if the surface S is regular.

(3)   The integral undergoes a discontinuity when passing through S such that

$$u\big|_{P_+} = 2\pi\tau(P) + \int_S \tau(Q)\frac{\cos(\mathbf{x}-\xi,\mathbf{n})}{|\mathbf{x}-\xi|^2}\,dS \tag{10}$$

and

$$u\big|_{P_-} = -2\pi\tau(P) + \int_S \tau(Q)\frac{\cos(\mathbf{x}-\xi,\mathbf{n})}{|\mathbf{x}-\xi|^2}\,dS \tag{11}$$

in the notation of the relations (7) and (8). Hence, the jump of u across S is

$$\tau = (1/4\pi)\left[u\big|_{P_+} - u\big|_{P_-}\right]. \tag{12}$$

(4)   The normal derivative remains continuous as S is crossed. The reader who is interested in the proof should look up the elegant proof given by Stakgold [19].

Interior and Exterior Dirichlet Problems

For the solution of a boundary value problem for an elliptic equation, we cannot prescribe u and $\partial u/\partial n$ arbitrarily on S. Therefore, equation (6) does not allow us to construct a solution for equation (3) such that u shall itself have arbitrary values on S and also arbitrary values for its normal derivative there. As such, there are two kinds of boundary value problems for elliptic equations. For one kind, we have the value of the solution prescribed on S—the so-called Dirichlet problem. For the second kind, the value of the normal derivative of the solution is prescribed on S—the Neumann problem. We discuss the Dirichlet problem first.

To fix the ideas, let us discuss the Dirichlet problem for the region exterior to the unit sphere. In order to get a unique solution, it is necessary to impose some sort of boundary condition at infinity along with the boundary value on the surface S of the sphere. Indeed, the functions $u_1(\mathbf{x}) = 1$ and $u_2(\mathbf{x}) = 1/r$ are both harmonic in the region R_e and assume the same value 1 on S. But if we require that the solution vanishes at infinity, then u_2 is the required solution. As a matter of fact, it is an important result in the potential theory [9, 19] that, when one solves the Dirichlet problem for the exterior of the unit sphere (by expansions in spherical harmonics) such that the potential vanishes at infinity, then one finds that the behavior of the solution is

$$u\Big|_\infty = O\left(\frac{1}{r}\right), \qquad \frac{\partial u}{\partial r}\Big|_\infty = O\left(\frac{1}{r^2}\right). \tag{13}$$

From these considerations and from the value of the fundamental solution, it is traditionally proved that

$$\lim_{r \to \infty} \int_S \left(E\frac{\partial u}{\partial r} - u\frac{\partial E}{\partial r} \right) dS = 0 \tag{14}$$

on the surface of the sphere of radius r.

We can now define and analyze the exterior and interior Dirichlet problems for an arbitrary surface S as follows.

Definition. The exterior Dirichlet problem is the boundary value problem

$$\nabla^2 u_e = 0 , \qquad \mathbf{x} \in R_e ; \qquad u_e\big|_S = f ,$$

$$u_e\big|_\infty = O\left(\frac{1}{r}\right), \qquad \frac{\partial u_e}{\partial r}\Big|_\infty = O\left(\frac{1}{r^2}\right), \tag{15}$$

where $f(\mathbf{x})$ is a given continuous function on S.

Definition. The interior Dirichlet problem is the boundary value problem

$$\nabla^2 u_i = 0 , \qquad \mathbf{x} \in R_i , \qquad u_i|_S = f . \tag{16}$$

Suppose that we are required to find the solution of the interior Dirichlet problem. We assume that such a solution u is the potential of a double layer with density τ (which is as yet unknown):

$$u_i(\mathbf{x}) = \int_S \frac{\tau(\xi) \cos(\mathbf{x} - \xi, \mathbf{n})}{r^2} \, dS . \tag{17}$$

For u_i to satisfy the boundary condition $(16)_2$ from within S, we appeal to the relation (11) and get the Fredholm integral equation of the second kind for $\tau(P)$:

$$\tau(P) = -(1/2\pi)f(P) + \int_S K(P, Q)\tau(Q) \, dS , \tag{18}$$

where the kernel $K(P, Q)$ is

$$K(P, Q) = \left[\cos(\mathbf{x} - \xi, \mathbf{n})\right]/2\pi \, |\mathbf{x} - \xi|^2 , \tag{19}$$

and $P \; (=\mathbf{x})$ and $Q \; (=\xi)$ are both on S. We solve the integral equation (18) for τ, substitute this solution in (17), and obtain the required solution of the boundary value problem (16).

In exactly the same way, the Dirichlet problem for an external domain bounded internally by S can be reduced to the solution of a Fredholm integral equation of the second kind.

We can present an integral-equation formulation of the exterior and interior boundary value problems (15) and (16) in a composite medium when $f(\mathbf{x})$ is the same function in both these problems. Recall that the fundamental solution $E(\mathbf{x}; \xi)$ satisfies the relation

$$-\nabla^2 E = \delta(\mathbf{x} - \xi) , \qquad \text{for all} \quad \mathbf{x} \quad \text{and} \quad \xi . \tag{20}$$

Multiply (16) by E, (20) by u_i, add, integrate and apply Green's second identity. This results in

$$\int_S \left(E \frac{\partial u_i}{\partial n} - u_i \frac{\partial E}{\partial n} \right) dS = \begin{cases} u_i(\xi) , & \xi \in R_i , \\ 0 , & \xi \in R_e , \end{cases} \tag{21}$$

where $\mathbf{n}$ is the outward normal to R_i on S.

The corresponding result for the exterior region is obtained by multiplying (15) by E, (20) by u_e, adding, integrating over the region bounded internally by S and externally by a sphere S_r, and applying Green's second identity. The contribution from S_r vanishes as $r \to \infty$ in view of the boundary condition at infinity, and we finally have

$$\int_S \left(-E\frac{\partial u_e}{\partial n} + u_e \frac{\partial E}{\partial n} \right) dS = \begin{cases} 0, & \xi \in R_i, \\ u_e(\xi), & \xi \in R_e, \end{cases} \tag{22}$$

where we have used the fact that the outward normal to R_e on S is in the $-\mathbf{n}$ direction.

The next step is to add (21) and (22) and observe that both u_i and u_e take the same value f as we approach the surface. Thus, we obtain

$$\int_S E(\mathbf{x}; \xi) \left(\frac{\partial u_i}{\partial n} - \frac{\partial u_e}{\partial n} \right) dS = \begin{cases} u_i(\xi), & \xi \in R_i, \\ u_e(\xi), & \xi \in R_e, \end{cases} \tag{23}$$

and $\mathbf{x} \in S$. Let us make use of the relations (2) and (9), and relabel $\mathbf{x}$ and ξ; thereby, we end up with the relation

$$\int_S \frac{\sigma(\xi)}{|\mathbf{x}-\xi|} dS = \begin{cases} u_i(\mathbf{x}), & \mathbf{x} \in R_i, \\ u_e(\mathbf{x}), & \mathbf{x} \in R_e, \end{cases} \tag{24}$$

that is, a single-layer potential with unknown charge density σ. Finally, using the boundary condition

$$u_i|_S = u_e|_S = f,$$

in (24), we obtain the Fredholm integral equation of the first kind

$$f(\mathbf{x}) = \int_S [\sigma(\xi)/|\mathbf{x}-\xi|] \, dS, \tag{25}$$

with both $\mathbf{x}$ and ξ on S.

Interior and Exterior Neumann Problems

In this case we are required to find the solution of Laplace or Poisson equation when the normal derivative is prescribed.

Definition. The exterior Neumann problem is the boundary value problem

$$\nabla^2 u_e = 0, \quad \mathbf{x} \in R_e, \quad \left.\frac{\partial u_e}{\partial n}\right|_S = f, \quad u_e\big|_\infty = 0. \quad (26)$$

Definition. The interior Neumann problem is the boundary value problem

$$\nabla^2 u_i = 0, \quad \mathbf{x} \in R_i, \quad \left.\frac{\partial u_i}{\partial n}\right|_S = f. \quad (27)$$

For a Neumann problem, the prescribed function $f(\mathbf{x})$ satisfies the consistency condition

$$\int_S f(\xi)\, dS = 0, \quad \xi \in S, \quad (28)$$

which follows by integrating the identity

$$\int_{R_i} (\nabla^2 u_i)\, dV = 0,$$

and using the divergence theorem.

The exterior and interior Neumann problems can be reduced to integral equations in a manner similar to the one explained for the corresponding Dirichlet problem. Indeed, we seek a solution of the interior Neumann problem in the form of the potential of a simple layer

$$u_i = \int_S \left[\sigma(Q)/r\right] dS, \quad (29)$$

which is a harmonic function in R_i. It will be a solution of (27) if the density σ is so chosen that

$$\left.\frac{\partial u_i}{\partial n}\right|_{P_-} = f(P), \quad P \in S. \quad (30)$$

Appealing to the relation (8), we have

$$f(P) = \left.\frac{\partial u_i}{\partial n}\right|_{P_-} = 2\pi\sigma(P) + \int_S \sigma(Q)\frac{\cos(\xi - \mathbf{x}, \mathbf{n})}{r^2}\, dS. \quad (31)$$

Thus, $\sigma(P)$ is a solution of the Fredholm integral equation of the second kind

$$\sigma(P) = \frac{1}{2\pi} f(P) - \int_S \sigma(Q) \frac{\cos(\boldsymbol{\xi} - \mathbf{x}, \mathbf{n})}{2\pi r^2} \, dS \, . \tag{32}$$

The solution of the exterior Neumann problem also leads to a similar integral equation. Furthermore, we can give the integral-equation formulation of the problems (26) and (27) in a composite medium when f is the same function in both these problems. Proceeding as we did for the corresponding Dirichlet problem, we obtain a Fredholm integral equation of the first kind. Instead of a single-layer potential, we now get a double-layer potential. The details are left to the reader.

Let us observe in passing that the solution of (27) is not unique, since an arbitrary constant can be added to a solution and the resulting function will satisfy (27).

6.3. EXAMPLES

Example 1.   *Electrostatic potential due to a thin circular disk.*   Let us take S to be a circular disk of radius a on which the potential V is prescribed. Let us choose cylindrical polar coordinates (ρ, φ, z) such that the origin is on the center of the disk with the z axis normal to the plane of the disk. Thus, the disk occupies the region $z = 0$, $0 \leqslant \rho \leqslant a$, for all φ. There is no loss of generality in taking the potential V on the disk as $f^{(n)}(\rho) \cos n\varphi$, where n is an arbitrary integer, because we can use the Fourier superposition principle. The charge density σ will also then have the form $\sigma^{(n)}(\rho) \cos n\varphi$. From (6.2.25), we have

$$f^{(n)}(\rho) \cos n\varphi = \int_{\text{disk}} [\sigma(\boldsymbol{\xi})/|\mathbf{x} - \boldsymbol{\xi}|] \, dS \, , \tag{1}$$

where $\mathbf{x} = (\rho, \varphi, 0)$ and $\boldsymbol{\xi} = (t, \varphi_1, 0)$. Or

$$f^{(n)}(\rho) \cos n\varphi = \int_0^a \int_0^{2\pi} \frac{t\sigma^{(n)}(t)(\cos n\varphi_1) \, d\varphi_1 \, dt}{[\rho^2 + t^2 - 2\rho t \cos(\varphi - \varphi_1)]^{\frac{1}{2}}} \, . \tag{2}$$

But, by setting $\varphi_1 - \varphi = \psi$, we find that

$$\int_0^{2\pi} \frac{\cos n\varphi_1 \, d\varphi_1}{[\rho^2 + t^2 - 2\rho t \cos(\varphi - \varphi_1)]^{\frac{1}{2}}} = \int_{-\varphi}^{2\pi - \varphi} \frac{\cos n(\varphi + \psi) \, d\psi}{[\rho^2 + t^2 - 2\rho t \cos(\varphi - \varphi_1)]^{\frac{1}{2}}}$$

$$= \int_{-\varphi}^{2\pi - \varphi} \frac{(\cos n\varphi)(\cos n\psi) \, d\psi}{[\rho^2 + t^2 - 2\rho t \cos \psi]^{\frac{1}{2}}}$$

$$= (\cos n\varphi) \left[\int_{-\varphi}^{0} + \int_{0}^{2\pi} + \int_{2\pi}^{2\pi - \varphi} \frac{\cos n\psi \, d\psi}{[\rho^2 + t^2 - 2\rho t \cos \psi]^{\frac{1}{2}}} \right]$$

$$= (\cos n\varphi) \int_0^{2\pi} \frac{\cos n\psi \, d\psi}{[\rho^2 + t^2 - 2\rho t \cos \psi]^{\frac{1}{2}}} , \tag{3}$$

where in the first integral we have put $\psi' = \psi + 2\pi$.

From (2) and (3), there follows the equation

$$f^{(n)}(\rho) = \int_0^a \int_0^{2\pi} \frac{t\sigma^{(n)}(t) \cos n\psi \, d\psi \, dt}{[\rho^2 + t^2 - 2\rho t \cos \psi]^{\frac{1}{2}}} . \tag{4}$$

Finally, we use in (4) the expansion formula

$$[\rho^2 + t^2 - 2\rho t \cos \psi]^{-\frac{1}{2}} = \sum_{r=0}^{\infty} \int_0^{\infty} (2 - \delta_{0r})(\cos r\psi) J_r(p\rho) J_r(pt) \, dp , \tag{5}$$

where δ_{0r} is the Kronecker delta, and use the orthogonality of the cosine function. The result is the Fredholm integral equation of the first kind

$$f^{(n)}(\rho) = \int_0^a t\sigma^{(n)}(t) K_0(t, \rho) \, dt , \tag{6}$$

where the kernel $K_0(t, \rho)$ is

$$K_0(t, \rho) = 2\pi \int_0^{\infty} J_n(p\rho) J_n(pt) \, dp . \tag{7}$$

For an annular disk of inner radius b and outer radius a, the formula that corresponds to (6) is

$$f^n(\rho) = \int_b^a t\sigma^n(t) K_0(t, \rho) \, dt \,. \tag{8}$$

Example 2. Solve the integral equation (6.2.25) when S is a unit sphere and $f = \sin\theta\cos\varphi$; that is, solve the integral equation

$$\sin\theta\cos\varphi = \int_0^{2\pi} d\varphi_1 \int_0^\pi \frac{(\sin\theta_1)\,\sigma(\theta_1, \varphi_1)\,d\theta_1}{|\mathbf{x} - \boldsymbol{\xi}|} \,. \tag{9}$$

Here, we use the expansion formula

$$\frac{1}{|\mathbf{x} - \boldsymbol{\xi}|} = \sum_{n=0}^\infty N_{0,n} \sum_{m=-n}^n \frac{Y_n^m(\theta, \varphi)\,Y_n^{*m}(\theta_1, \varphi_1)}{N_{m,n}} \,, \tag{10}$$

where $Y_n^m(\theta, \varphi)$ are the spherical harmonics and

$$N_{m,n} = \int_0^{2\pi} d\varphi \int_0^\pi (\sin\theta)\,|Y_n^m(\theta, \varphi)|^2 \, d\theta$$

$$= \frac{4\pi}{2n+1} \frac{(n+|m|)!}{(n-|m|)!} \,. \tag{11}$$

Furthermore, we set

$$\sigma(\theta_1, \varphi_1) = \sum_{m=0}^\infty \sum_{m=-n}^n \sigma_{m,n} Y_n^m(\theta_1, \varphi_1) \,, \tag{12}$$

and note that $\sin\theta\cos\varphi = \frac{1}{2}[Y_1^1(\theta, \varphi) + Y_1^{-1}(\theta, \varphi)]$. Putting (10)–(12) in (9) and using the orthogonality properties of the spherical harmonics, we obtain

$$\sigma_{1,1} = \sigma_{1,-1} = 3/8\pi \,,$$

and $\sigma_{m,n} \equiv 0$ for all other m and n. Thus, from (12), it follows that

$$\sigma(\theta, \varphi) = (3/4\pi) P_1^1(\cos\theta)\cos\varphi \,, \tag{13}$$

where P_1^1 is an associated Legendre function.

6.4. GREEN'S FUNCTION APPROACH

The Green's function is an auxiliary function which plays the same crucial role in the integral-equation formulation of partial differential equations as it plays in the case of ordinary differential equations. This function depends on the form of the differential equation, the boundary condition, and the region. For instance, the Green's function $G(\mathbf{x};\xi)$ for the Laplace equation in an open, bounded region R in three-dimensional space with boundary S is the solution of the boundary value problem

$$-\nabla^2 G = \delta(\mathbf{x}-\xi) , \qquad G|_S = 0 , \tag{1}$$

where $\mathbf{x}$ and ξ are in R. In the language of electrostatics, the function G is the electrostatic potential due to a unit charge at ξ when the surface S is a grounded metallic shell. As such, G is the sum of the potential of the unit source at ξ in free space and the potential due to the charge induced on S:

$$G(\mathbf{x};\xi) = (1/4\pi\,|\mathbf{x}-\xi|) + v(\mathbf{x};\xi) , \tag{2}$$

where v is a harmonic function which satisfies the boundary value problem

$$\nabla^2 v = 0 , \qquad \mathbf{x} \in R , \qquad v|_S = -E . \tag{3}$$

Let us show that the Green's function is symmetric. When $G(\mathbf{x};\xi)$ and $G(\mathbf{x};\eta)$ are the Green's functions for the region R corresponding to the sources at ξ and η, we have the relations

$$-\nabla^2 G(\mathbf{x};\xi) = \delta(\mathbf{x}-\xi) , \qquad G|_S = 0 , \tag{4}$$

$$-\nabla^2 G(\mathbf{x};\eta) = \delta(\mathbf{x}-\eta) , \qquad G|_S = 0 . \tag{5}$$

The result of the routine steps of multiplying (4) by $G(\mathbf{x};\eta)$, (5) by $G(\mathbf{x};\xi)$, subtracting, integrating, and applying Green's second identity is

$$\int_S \left[G(\mathbf{x};\xi)\frac{\partial G(\mathbf{x};\eta)}{\partial n} - G(\mathbf{x};\eta)\frac{\partial G(\mathbf{x};\xi)}{\partial n} \right] dS = G(\xi;\eta) - G(\eta;\xi). \tag{6}$$

The symmetry of the Green's function follows by applying the boundary conditions in (6).

Note also that the fundamental solution $E(\mathbf{x};\xi)$ is the free-space Green's function.

Solution of the Dirichlet Problem

We are now ready to give an integral-equation formulation to the boundary value problem

$$-\nabla^2 u(\mathbf{x}) = 4\pi\rho(\mathbf{x}), \qquad \mathbf{x} \in R, \qquad u|_S = f, \tag{7}$$

in terms of the Green's function. For this purpose, multiply (1) by u and (7) by G, subtract, integrate, use Green's second identity, and get

$$u(\xi) = 4\pi \int_R G(\mathbf{x};\xi)\rho(\mathbf{x})\, dV - \int_S f(\mathbf{x})\,[\partial G(\mathbf{x};\xi)/\partial n]\, dS\,. \tag{8}$$

By interchanging $\mathbf{x}$ and ξ and using the symmetry of the Green's function, we find the representation formula

$$u(\mathbf{x}) = 4\pi \int_R G(\mathbf{x};\xi)\rho(\xi)\, dV - \int_S [\partial G(\mathbf{x};\xi)/\partial n]f(\xi)\, dS\,. \tag{9}$$

For the particular case $\rho \equiv 0$, the formula (9) becomes

$$u(\mathbf{x}) = -\int_S f(\xi)\,[\partial G(\mathbf{x};\xi)/\partial n]\, dS\,. \tag{10}$$

When $f = 1$ on S, then the solution u of the Laplace equation is clearly $u = 1$ for the interior Dirichlet problem. Thus, (10) yields an interesting relation,

$$-\int_S [\partial G(\mathbf{x};\xi)]/\partial n]\, dS = 1\,, \qquad \mathbf{x} \in R\,. \tag{11}$$

Example. *Poisson integral formula.* The Green's function for the Laplace equation, when the surface S is a sphere, can be found by various methods. The easiest method is to express it as source and image point combination. Let the radius of the sphere be a (see Figure 6.1).

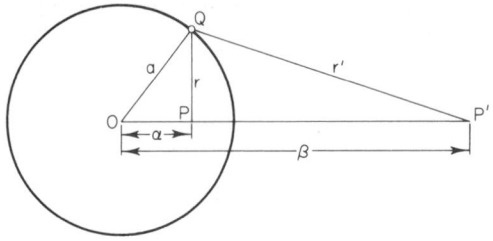

Figure 6.1

For any point P ($=\mathbf{x}$) with radial distance α within the sphere, we have an inverse point P' ($=\mathbf{x}'$) on the same radial line at a radial distance β outside the sphere, such that $\alpha\beta = a^2$. If Q ($=\xi$) is any point on S, then the triangles OQP and OQP' are easily seen to be similar. Therefore, $r'/r = a/\alpha$, or

$$1/r = a/\alpha r' . \tag{12}$$

Examining the relations (1) and (2), we readily find the value of the Green's function to be

$$G(P, Q) = \frac{1}{4\pi}\left(\frac{1}{r} - \frac{a}{\alpha}\frac{1}{r'}\right). \tag{13}$$

Having found the Green's function, we can solve the interior Dirichlet problem for the sphere:

$$\nabla^2 u = 0, \quad r < a; \quad u = f(\theta, \varphi) \quad \text{on} \quad r = a. \tag{14}$$

To use the formula (10) we need the value of $\partial G/\partial n$. This is obtained if we observe from the figure that

$$\alpha^2 = a^2 + r^2 - 2ar\cos(\mathbf{x}-\xi, \mathbf{n}) ;$$
$$\beta^2 = a^2 + r'^2 - 2ar'\cos(\mathbf{x}'-\xi, \mathbf{n}) . \tag{15}$$

Thus,

$$\frac{\partial G}{\partial n} = \frac{1}{4\pi}\left(-\frac{1}{r^2}\frac{\partial r}{\partial n} + \frac{a}{\alpha}\frac{1}{r'^2}\frac{\partial r'}{\partial n}\right) = \frac{1}{4\pi}\left(\frac{\cos(\mathbf{x}-\boldsymbol{\xi},\mathbf{n})}{r^2} - \frac{a}{\alpha}\frac{\cos(\mathbf{x}'-\boldsymbol{\xi},\mathbf{n})}{r'^2}\right)$$

$$= \frac{1}{4\pi}\left(\frac{\alpha^2-a^2-r^2}{2ar^3} - \frac{a}{\alpha}\frac{\beta^2-a^2-r'^2 t}{2ar'^3}\right) = -\frac{a^2-\alpha^2}{4\pi ar^3}, \tag{16}$$

where we have used the relations $r'/r = a/\alpha$ and $\alpha\beta = a^2$.

Substituting (16) in (10), we finally have

$$u(P) = \frac{a(a^2-\alpha^2)}{4\pi}\int\limits_0^\pi\int\limits_0^{2\pi}\frac{f(\theta',\varphi')(\sin\theta')\,d\theta'\,d\varphi'}{(\alpha^2+a^2-2a\alpha\cos(\mathbf{x},\boldsymbol{\xi})^{3/2}}. \tag{17}$$

Neumann Problem

By defining the Green's function $G(\mathbf{x};\boldsymbol{\xi})$ by the boundary value problem

$$-\nabla^2 G(\mathbf{x};\boldsymbol{\xi}) = \delta(\mathbf{x}-\boldsymbol{\xi}), \qquad \partial G/\partial n\big|_s = 0, \tag{18}$$

we can extend the above analysis to the Neumann problem. Indeed, the integral equation that corresponds to (10) for the Neumann problem

$$-\nabla^2 u = 0, \qquad \mathbf{x} \in R, \qquad \partial u/\partial n = f \tag{19}$$

is

$$u(\mathbf{x}) = \int\limits_S G(\mathbf{x};\boldsymbol{\xi})f(\boldsymbol{\xi})\,dS. \tag{20}$$

Of course, the function f must satisfy the consistency condition (6.2.28).

Finally, let us consider the interior and exterior Dirichlet problems for a body S enclosed within a surface Σ:

$$\nabla^2 u_i = 0, \qquad \mathbf{x} \in R_i, \qquad u_i\big|_s = f, \tag{21}$$

$$\nabla^2 u_e = 0, \qquad \mathbf{x} \in R_e, \qquad u_e\big|_s = f, \qquad u_e\big|_\Sigma = 0. \tag{22}$$

The Green's function $G(\mathbf{x};\boldsymbol{\xi})$ satisfies the auxiliary problem (we absorb the factor 4π in G):

$$-\nabla^2 G = 4\pi\delta(\mathbf{x}-\boldsymbol{\xi}), \qquad G\big|_\Sigma = 0. \tag{23}$$

We now follow the same steps as we did in deriving the formula (6.2.25). Thus, from the relations (21)–(23), we obtain

$$f(\mathbf{x}) = \int_S G(\mathbf{x};\xi)\sigma(\xi)\,dS\,, \tag{24}$$

which reduces to (6.2.25) for an unbounded medium.

6.5. EXAMPLES

Example 1. *Electrostatic potential problem of a conducting disk bounded by two parallel planes.* This problem is an extension of the problem considered in Example 1 of Section 6.3. We follow that notation and assume that the parallel planes are $z = b$ and $z = -c$ $(b, c > 0)$. The boundary value problem becomes

$$\nabla^2 V(\rho, \varphi, z) = 0 \quad \text{in} \quad D\,, \tag{1}$$

$$V(\rho, \varphi, 0) = f^{(n)}(\rho)\cos n\varphi\,, \quad 0 \leqslant \rho \leqslant a\,, \tag{2}$$

$$V(\rho, \varphi, z) = 0\,, \quad z = b\,, \quad z = -c\,, \tag{3}$$

where D is the region between the disk and the parallel planes.

The Green's function G corresponding to this problem satisfies the auxiliary system

$$-\nabla^2 G(\mathbf{x};\xi) = 4\pi\delta(\mathbf{x}-\xi)\,, \quad G = 0 \quad \text{on} \quad z = b\,, \quad z = -c\,. \tag{4}$$

This function is found easily by the method of images. Indeed, for a positive unit charge at the source point $\xi = (t, \varphi_1, z_1)$, the image system consists of positive unit charge at the points (see Figure 6.2)

Figure 6.2

$$\xi_n^+ = [t, \varphi_1, 2n(b+c) + z_1], \qquad n = \pm 1, \pm 2, \ldots, \tag{5}$$

and negative unit charge at points

$$\xi_n^- = [t, \varphi_1, 2n(b+c) - 2c - z_1], \qquad n = 0, \pm 1, \pm 2, \ldots. \tag{6}$$

The value of the Green's function therefore is

$$G(\mathbf{x};\xi) = \frac{1}{|\mathbf{x}-\xi|} + \sum_{n=1}^{\infty} \frac{1}{|\mathbf{x}-\xi_n^+|} + \sum_{n=-1}^{-\infty} \frac{1}{|\mathbf{x}-\xi_n^+|} - \sum_{n=-\infty}^{\infty} \frac{1}{|\mathbf{x}-\xi_n^-|}. \tag{7}$$

The next step is to use the identity

$$1/|\mathbf{x}-\xi| = \int_0^{\infty} J_0(p\varpi)(\exp -p\,|z-z_1|)\,dp, \tag{8}$$

where $\varpi = [\rho^2 + t^2 - 2\rho t \cos(\varphi - \varphi_1)]^{1/2}$. Then, the relation (7) takes the form

$$G(\mathbf{x};\xi) = (1/|\mathbf{x}-\xi|) + \int_0^{\infty} J_0(p\varpi)\Big[\sum_1^{\infty} (\exp -p\,|z - 2n(b+c) - z_1|)$$

$$+ \sum_{-1}^{-\infty} (\exp -p\,|z - 2n(b+c) - z_1|)$$

$$- \sum_0^{\infty} (\exp -p\,|z - 2n(b+c) + 2c + z_1|)$$

$$- \sum_{-1}^{-\infty} (\exp -p\,|z - 2n(b+c) + 2c + z_1|)\Big]\,dp. \tag{9}$$

After summing the geometric series which occur in relation (9) and slight simplification, we obtain

$$G(\mathbf{x};\xi) = \frac{1}{|\mathbf{x}-\xi|} + \int_0^{\infty} J_0(p\varpi)$$

$$\times \frac{e^{-(c+z_1)p}[\sinh p(z-b)] - e^{-(n-z_1)p}[\sinh p(z+c)]}{\sinh p(b+c)}\,dp. \tag{10}$$

Finally, the result of using the expansion of $J_0(p\varpi)$,

$$J_0(p\varpi) = \sum_{r=0}^{\infty} (2-\delta_{0r})[\cos r(\varphi - \varphi_1)]J_r(p\rho)J_r(pt), \qquad (11)$$

in (10) is

$$G(\mathbf{x};\boldsymbol{\xi}) = \frac{1}{|\mathbf{x}-\boldsymbol{\xi}|} + \sum_{r=0}^{\infty} (2-\delta_{0r})[\cos r(\varphi - \varphi_1)]G^{(r)}(\rho, t, z, z_1), \qquad (12)$$

where

$$G^{(r)}(\rho, t, z, z_1) = \int_0^{\infty} \frac{e^{-p(c+z_1)}[\sinh p(z-b)] - e^{-p(b-z_1)}[\sinh p(z+c)]}{\sinh p(b+c)}$$

$$\times J_r(p\rho)J_r(pt)\, dp . \qquad (13)$$

To derive the integral equation, we multiply (1) by G and (4) by V, follow the routine procedure, and get

$$V(\rho, \varphi, z) = \frac{1}{4\pi} \int_{S^+ + S^-} \left(G\frac{\partial V}{\partial n} - V\frac{\partial G}{\partial n} \right) dS , \qquad (14)$$

where S^+ and S^- are the upper and lower parts of the disk, respectively. On the surfaces $S^\pm$, the value of the outward normal is $\mp \partial/\partial z_1$, respectively. Using this fact and the boundary conditions (2) in (14), we have

$$f^{(n)}(\rho)\cos n\varphi = \int_0^a \int_0^{2\pi} t\sigma^{(n)}(t)\, G(\rho, t, \varphi, \varphi_1, 0, 0)(\cos n\varphi_1)\, dp_1\, dt , \qquad (15)$$

where $\sigma^{(n)}(t)\cos n\varphi_1 = (1/4\pi)(\partial V/\partial z_{1+} - \partial V/\partial z_{1-})$. Setting $\varphi_1 - \varphi = \psi$, and following the steps which led from (6.3.2) to (6.3.4), we obtain

$$f^{(n)}(\rho) = \int_0^a t\sigma^{(n)}(t)\, dt \int_0^{2\pi} \frac{\cos n\psi \, d\psi}{|\mathbf{x}-\boldsymbol{\xi}|_{z=z_1=0}}$$

$$+ \int_0^a t\sigma^{(n)}(t)[2\pi G^{(n)}(\rho, t, 0, 0)]\, dt , \qquad (16)$$

where we have substituted the value of G as in (12). This can be written as

$$f''(\rho) = \int_0^a t\sigma^{(n)}(t) K_1(t,\rho)\, dt\,, \tag{17}$$

where

$$K_1(t,\rho) = \int_0^{2\pi} \frac{\cos n\psi \, d\psi}{[\rho^2 + t^2 - 2\rho t \cos\psi]^{\frac{1}{2}}} + 2\pi G^{(n)}(\rho,t,0,0)\,. \tag{18}$$

When $b \to \infty$, $c \to \infty$, we recover the formula (6.3.4).

For an annular disk of inner radius b and outer radius a, the formula which corresponds to (17) is

$$f^{(n)}(\rho) = \int_b^a t\sigma^{(n)}(t) K_1(t,\rho)\, dt \tag{19}$$

with the same kernel as in (18).

Example 2. *Electrostatic potential problem of an axially symmetric conductor placed symmetrically inside a cylinder of radius b.* We again take cylindrical polar coordinates (ρ, φ, z) with origin at the center of the conductor and z axis along its axis of symmetry (which is also the axis of the cylinder). For simplicity, we take V on the surface S of the conductor to be unity. Then, from the relation (6.4.24), we have the Fredholm integral equation of the first kind

$$1 = \int_S G(\mathbf{x};\boldsymbol{\xi})\sigma(\boldsymbol{\xi})\, dS\,, \qquad \mathbf{x},\boldsymbol{\xi} \in S\,, \tag{20}$$

where G satisfies the system

$$-\nabla^2 G = 4\pi\delta(\mathbf{x}-\boldsymbol{\xi})\,, \qquad G = 0 \quad \text{on} \quad \rho = b\,. \tag{21}$$

In terms of cylindrical polar coordinates, the differential equation for G becomes

$$\frac{1}{\rho}\left(\rho\frac{\partial G}{\partial\rho}\right) + \frac{1}{\rho^2}\frac{\partial^2 G}{\partial\varphi^2} + \frac{\partial^2 G}{\partial z^2} = -\frac{4\pi}{\rho}\delta(\rho-t)\delta(\psi)\delta(z-z_1)\,, \tag{22}$$

where $\varphi_1 - \varphi = \psi$. From the definition (6.4.2) of the Green's function, we know that

$$G_1(\mathbf{x};\boldsymbol{\xi}) = G(\mathbf{x};\boldsymbol{\xi}) - (1/|\mathbf{x}-\boldsymbol{\xi}|) \tag{23}$$

is finite in the limit as $\mathbf{x} \to \boldsymbol{\xi}$. We can calculate the solution of (22) by taking the Fourier series expansion

$$G(\mathbf{x};\boldsymbol{\xi}) = \sum_{r=1}^{\infty} (2 - \delta_{0r})(\cos r\psi) \, g^{(r)}(\rho, t, z, z_1) \,, \tag{24}$$

where

$$g^{(r)} = (1/2\pi) \int_{0}^{2\pi} G(\mathbf{x};\boldsymbol{\xi})(\cos r\psi) \, d\psi \,.$$

Multiply the differential equation (22) by $(1/2\pi)\cos r\psi$ and integrate with respect to ψ from 0 to 2π. The result is

$$\frac{1}{\rho}\frac{\partial}{\partial\rho}\left(\rho\frac{\partial}{\partial\rho} g^{(r)}\right) - \frac{r^2}{\rho^2} g^{(r)} + \frac{\partial^2 g^{(r)}}{\partial z^2} = -\frac{2}{\rho}\delta(\rho - \rho_1)\delta(z - z_1) \,. \tag{25}$$

Next, we take the Fourier transform of equation (25) by setting

$$T(g^{(r)}) = (2\pi)^{-\frac{1}{2}} \int_{-\infty}^{\infty} e^{ipz} g^{(r)} \, dz \,;$$

$$g^{(r)} = (2\pi)^{-\frac{1}{2}} \int_{-\infty}^{\infty} e^{-ipz} T(g^{(r)}) \, dp \,. \tag{26}$$

The system (21) becomes

$$\rho^2 \frac{d^2}{d\rho^2} T(g^{(r)}) + \rho \frac{d}{d\rho} T(g^{(r)}) - (\rho^2 p^2 + r^2) T(g^{(r)})$$

$$= -\frac{\sqrt{2}}{\rho\sqrt{\pi}} e^{ipz_1} \delta(\rho - \rho_1) \,, \qquad T(g^{(r)}) = 0 \,, \qquad \rho = b \,. \tag{27}$$

This boundary value problem can be easily solved by the method and notation of Chapter 5 and the solution so obtained is then inverted to yield

$$g^{(r)} = (1/\pi) \int_{-\infty}^{\infty} e^{ip(z_1 - z)} \{K_r(p\rho_>) I_r(p\rho_<)$$

$$- [K_r(pb)/I_r(pb)] I_r(p\rho) I_r(pt)\} \, dp \,, \tag{28}$$

where I_r and K_r are modified Bessel functions. Finally, from (24) and (28), we find the value of G:

$$G(\mathbf{x};\boldsymbol{\xi}) = (1/\pi) \sum_{r=0}^{\infty} (2-\delta_{0r})(\cos r\psi) \int_{-\infty}^{\infty} e^{ip(z_1-z)}$$

$$\times \{K_r(p\rho_>)I_r(p\rho_<) - [K_r(pb)/I_r(pb)]I_r(p\rho)I_r(pt)\}\,dp\,. \quad (29)$$

When $b \to \infty$, $G = 1/|\mathbf{x}-\boldsymbol{\xi}|$, and we have from (29),

$$1/|\mathbf{x}-\boldsymbol{\xi}| = (1/\pi) \sum_{r=0}^{\infty} (2-\delta_{0r})(\cos r\psi) \int_{-\infty}^{\infty} e^{ip(z_1-z)} K_r(p\rho_>)I_r(p\rho_<)\,dp\,. \quad (30)$$

Combining (29) and (30), we have

$$G(\mathbf{x};\boldsymbol{\xi}) = (1/|\mathbf{x}-\boldsymbol{\xi}|) + \sum_{r=0}^{\infty} (2-\delta_{0r})(\cos r\psi)\,G^{(r)}(\rho,t,z,z_1)\,, \quad (31)$$

where

$$G^{(r)}(\rho,t,z,z_1) = -(1/\pi) \int_{-\infty}^{\infty} e^{-ip(z-z_1)} I_r(p\rho)I_r(pt)[K_r(pb)/I_r(pb)]\,dp$$

$$= -(1/\pi)\left[\int_{0}^{\infty} e^{-ip(z-z_1)} I_r(p\rho)I_r(pt)[K_r(pb)/I_r(pb)]\,dp\right.$$

$$\left. + \int_{-\infty}^{-0} e^{-ip(z-z_1)} I_r(p\rho)I_r(pt)[K_r(pb)/I_r(pb)]\,dp\right]. \quad (32)$$

Changing p to $-p$ in the second integral and observing that

$$I_r(-z) = (-1)^r I_r(z)\,, \qquad K_r(-z) = (-1)^r K_r(z)\,,$$

and hence $K_r(-pb)/I_r(-pb) = K_r(pb)/I_r(pb)$, we have from (32)

$$G^{(r)}(\rho,t,z,z_1) = -(2/\pi)\int_{0}^{\infty}\left[\frac{I_r(p\rho)I_r(pt)K_r(pb)}{I_r(pb)}\right]\cos[p(z-z_1)]\,dp. \quad (33)$$

So far, we have not used the fact that the conductor is axially symmetric. For an axially symmetric body, the Green's function is independent of φ. That leaves only one term in the series (31):

$$G(\mathbf{x};\boldsymbol{\xi}) = (1/|\mathbf{x}-\boldsymbol{\xi}|) + G^{(0)}(\rho,t,z,z_1)\,. \quad (34)$$

Equation (34) is of the form (23). Substituting (34) in (20), we have the required integral equation. When $b \to \infty$, we recover equation (6.3.4) for $f^{(n)}(\rho) = 1$.

6.6. THE HELMHOLTZ EQUATION

The discussion of the previous two sections can be easily extended to the case of the Helmholtz equation

$$(\nabla^2 + \lambda) u = 0 . \tag{1}$$

The free-space Green's function or the fundamental solution $E(\mathbf{x}; \xi)$ is the solution of the spherically symmetric differential equation

$$-\nabla^2 E - \lambda E = \delta(\mathbf{x} - \xi) , \tag{2}$$

and which vanishes at infinity. Such a solution in three dimensions is

$$E(\mathbf{x}; \xi) = \frac{\exp(i\,|\mathbf{x} - \xi|\,\sqrt{\lambda})}{4\pi\,|\mathbf{x} - \xi|} = \frac{\exp(ir\,\sqrt{\lambda})}{4\pi r} . \tag{3}$$

(i) When λ is a complex number, then $\sqrt{\lambda}$ is selected to be that root of λ that has a positive imaginary part so that E vanishes exponentially at infinity.

(ii) When λ is real and positive, that is, $\lambda = \omega^2$, ω real, the solution

$$E(\mathbf{x}; \xi) = \frac{\exp i\omega\,|\mathbf{x} - \xi|}{4\pi\,|\mathbf{x} - \xi|} = \frac{\exp i\omega r}{4\pi r} \tag{4}$$

is selected such that $\sqrt{\lambda} = \omega > 0$. This represents an outgoing wave if we adjoin the factor $e^{-i\omega t}$.

(iii) When λ is real and negative, we again choose $\sqrt{\lambda}$ in (3) to be the square root of λ that has a positive imaginary part for the same reason as in (i). For the particular case $\lambda = -k^2$, where k is real and positive, the formula (3) becomes

$$E(\mathbf{x}; \xi) = e^{-kr}/4\pi r . \tag{5}$$

The solutions which correspond to (3), (4), and (5) in two dimensions are

$$(i/4) H_0^{(1)}(|\mathbf{x}-\boldsymbol{\xi}|\sqrt{\lambda}), \qquad (i/4) H_0^{(1)}(\omega|\mathbf{x}-\boldsymbol{\xi}|), \qquad (1/2\pi) K_0(k|\mathbf{x}-\boldsymbol{\xi}|),$$

respectively. Here, the functions $H_0^{(1)}$ and K_0 are the Hankel and modified Bessel functions, respectively.

The integral representation formula for the solution of the inhomogeneous equation

$$(\nabla^2 - k^2)u = -4\pi\rho \tag{6}$$

is obtained from the relations (2) and (6) by using Green's identity and is readily found to be

$$u(\mathbf{x}) = \int_R \frac{\rho e^{-kr}}{r}\, dV + \frac{1}{4\pi}\int_S u\frac{\partial}{\partial n}\left(\frac{e^{-kr}}{r}\right) dS - \frac{1}{4\pi}\int_S \frac{e^{-kr}}{r}\frac{\partial u}{\partial n}\, dS . \tag{7}$$

The interpretation of these integrals as volume, single-layer, and double-layer potentials is the same as for the corresponding formulas in Section 6.2. The properties of these potentials are also similar. For instance, the formulas that correspond to (6.2.7) and (6.2.8) are

$$\left.\frac{\partial u}{\partial n}\right|_{P_\pm} = \mp 2\pi\sigma(P) + \int_S \sigma(Q)\frac{\partial}{\partial n}\left(\frac{e^{-kr}}{r}\right) dS , \tag{8}$$

where

$$u = \int_S \sigma(Q)\left(\frac{e^{-kr}}{r}\right) dS . \tag{9}$$

Similarly, the formulas that correspond to (6.2.10) and (6.2.11) are

$$u\big|_{P_\pm} = \pm 2\pi\tau(P) + \int_S \tau(Q)\frac{\partial}{\partial n}\,(e^{-kr}/r)\, dS , \tag{10}$$

where

$$u = \int_S \tau(Q)\frac{\partial}{\partial n}\,(e^{-kr}/r)\, dS . \tag{11}$$

The rest of the notation is the same as in Section 6.2.

The integral representation of solutions of the exterior and interior Dirichlet and Neumann problems is achieved in an analogous manner, as shall become evident from the examples of physical interest which are presented in the next section. We end this section by mentioning the so-called Sommerfeld radiation condition. A three-dimensional solution of the Helmholtz equation $(\nabla^2 + k^2)u = 0$ is said to satisfy the radiation condition if

$$\lim r\left(\frac{\partial u}{\partial r} - iku\right) = 0 , \tag{12}$$

as $r \to \infty$. Physically, this condition implies that there are no incoming waves from infinity. In two dimensions, the corresponding condition is

$$\lim \sqrt{r}\left(\frac{\partial u}{\partial r} - iku\right) = 0 , \tag{13}$$

as $r \to \infty$. The free-space Green's function satisfies the radiation condition.

6.7. EXAMPLES

Example 1. *Acoustic diffraction of a plane wave by a perfectly soft disk.* We follow the coordinate system and notation of Example 1 in Section 6.3. Furthermore, we assume the time dependence of the form $e^{-i\omega t}$ for the wave functions involved in the problem and omit this factor in the sequel. The time-independent part of the velocity potential u is

$$u(\rho, \varphi, z) = u_i(\rho, \varphi, z) + u_s(\rho, \varphi, z) , \tag{1}$$

where u_i and u_s denote the velocity potentials of the incident and diffracted fields. All three functions occurring in equation (1) satisfy the Helmholtz equation. The boundary value problem is

$$(\nabla^2 + k^2)u_s = 0 , \tag{2}$$

$$u_i(\rho \ \varphi, 0) + u_s(\rho, \varphi, 0) = 0 , \qquad 0 \leqslant \rho \leqslant a , \tag{3}$$

u_s and $\partial u_s/\partial z$ are continuous across $z = 0 , \qquad a < \rho < \infty , \tag{4}$

and u_s satisfies the radiation condition at infinity. The fundamental solution E that satisfies the equation

$$-(\nabla^2 + k^2)E = \delta(\mathbf{x} - \boldsymbol{\xi}) \tag{5}$$

as well as the radiation condition is

$$E(\mathbf{x};\boldsymbol{\xi}) = \frac{\exp(ik\,|\mathbf{x}-\boldsymbol{\xi}|)}{4\pi\,|\mathbf{x}-\boldsymbol{\xi}|}$$

$$= \frac{\exp ik\,\{\rho^2 + t^2 - 2\rho t[\cos(\varphi-\varphi_1)] + (z-z_1)^2\}^{\frac{1}{2}}}{4\pi\,\{\rho^2 + t^2 - 2\rho t[\cos(\varphi-\varphi_1)] + (z-z_1)^2\}^{\frac{1}{2}}}\,. \quad (6)$$

Multiply (2) by E, (5) by u_s, subtract, integrate, use Green's second identity, and obtain

$$u_s(\mathbf{x}) = \int_{S^+ + S^-} \left(E\frac{\partial u_s}{\partial n} - u_s\frac{\partial E}{\partial n} \right) dS\,, \quad (7)$$

where S^+ and S^- are the upper and lower surfaces of the disk, respectively. On $S^\pm$, we have $\partial/\partial n = \mp\partial/\partial z$, respectively. Thus, (7) can be written as

$$u_s(\rho,\varphi,z) = \int_0^a\int_0^{2\pi}\left[-E\frac{\partial u_s}{\partial z_1} + u_s\frac{\partial E}{\partial z_1}\right]_{z_1=0+} t\,d\varphi_1\,dt$$

$$+ \int_0^a\int_0^{2\pi}\left[E\frac{\partial u_s}{\partial z_1} - u_s\frac{\partial E}{\partial z_1}\right]_{z_1=0-} t\,d\varphi_1\,dt$$

$$= -\int_0^a\int_0^{2\pi} t\sigma(t,\varphi_1,0)\,E\big|_{z_1=0}\,d\varphi_1\,dt\,, \quad (8)$$

where we have used the fact that $u_s = -u_i$ on both sides of the disk and where

$$\sigma(t,\varphi_1,0) = \left(\frac{\partial u_s}{\partial z}\bigg|_{z_1=0+} - \frac{\partial u_s}{\partial z}\bigg|_{z_1=0-}\right). \quad (9)$$

When we apply the boundary condition (3) in (8), we get

$$u_i(\rho,\varphi,0) = \frac{1}{4\pi}\int_0^a\int_0^{2\pi}\frac{t\sigma(t,\varphi_1,0)\exp ik\,[\rho^2+t^2-2\rho t\cos(\varphi-\varphi_1)]^{\frac{1}{2}}}{[\rho^2+t^2-2\rho t\cos(\varphi-\varphi_1)]^{\frac{1}{2}}}$$

$$\times\,d\varphi_1\,dt\,. \quad (10)$$

In view of the Fourier superposition principle, we can assume that $u_i(\rho, \varphi, z)$ and $\sigma(\rho, \varphi)$ are of the form $u_i^{(n)}(\rho, z)\cos n\varphi$ and $2\sigma^{(n)}(\rho)\cos n\varphi$, respectively. Then, proceeding as in the steps that led from (6.3.2) to (6.3.4), we obtain from (10) the integral equation

$$u_i^{(n)}(\rho, 0) = \int_0^a t\sigma^{(n)}(t)\, K_1(t, \rho)\, dt \,, \tag{11}$$

where the kernel $K_1(t, \rho)$ is

$$K_1(t, \rho) = \frac{1}{2\pi} \int_0^{2\pi} \frac{\exp ik(\rho^2 + t^2 - 2\rho\cos\psi)^{1/2}}{(\rho^2 + t^2 - 2\rho t\cos\psi)^{1/2}} \cos n\psi\, d\psi \,. \tag{12}$$

The integral equation (11) is the required Fredholm integral equation of the first kind which embodies the solution of the boundary value problem (2)–(4).

For an annular disk with inner radius b and outer radius a, the corresponding integral equation is

$$u_i^{(n)}(\rho, 0) = \int_b^a t\sigma^{(n)}(t)\, K_1(t, \rho)\, dt \,. \tag{13}$$

Example 2.   *Torsional oscillations of an elastic half-space.* In terms of cylindrical polar coordinates, the axially symmetric boundary value problem

$$\frac{\partial^2 v}{\partial\rho^2} + \frac{1}{\rho}\frac{\partial v}{\partial\rho} - \frac{v}{\rho^2} + \frac{\partial^2 v}{\partial z^2} + k^2 v = 0 \,, \qquad k^2 = \frac{\omega^2 da^2}{\mu} \,, \tag{14}$$

$$v = \Omega\rho \,, \qquad z = 0 \,, \qquad 0 \leqslant \rho \leqslant 1 \,, \tag{15}$$

$$\partial v/\partial z = 0 \,, \qquad z = 0 \,, \qquad \rho > 1 \,, \tag{16}$$

embodies the torsional oscillations of a homogeneous and isotropic half-space which occupies the region $z \geqslant 0$. A disk of radius a is attached to it and is forced to execute torsional oscillations with period $2\pi/\omega$. All lengths are made dimensionless with a as the standard length, so that the disk occupies the region $z = 0$, $0 \leqslant \rho \leqslant 1$. The quantities d and μ are respectively the density and shear modulus of the elastic material, while k is a dimensionless parameter.

It is easily verified from (14) that the function $w(\rho, \varphi, z)$,

$$w(\rho, \varphi, z) = v(\rho, z)\cos\varphi \, , \tag{17}$$

satisfies the Helmholtz equation

$$(\nabla^2 + k^2)w = 0 \, . \tag{18}$$

The Green's function that corresponds to this boundary value problem satisfies the auxiliary system

$$(\nabla^2 + k^2)\,G(\mathbf{x};\xi) = -4\pi\delta(\mathbf{x}-\xi) \, ; \qquad \left.\frac{\partial G}{\partial z_1}\right|_{z_1=0} = 0 \, , \tag{19}$$

where, as before, $\mathbf{x} = (\rho, \varphi, z)$ and $\xi = (t, \varphi_1, z_1)$. Again the method of images gives us the value of G

$$G(\mathbf{x};\xi) = \frac{\exp ikr}{r} + \frac{\exp ikr'}{r'} \, , \tag{20}$$

where $r = |\mathbf{x}-\xi|$, $r' = |\mathbf{x}-\xi'|$, and ξ' is the image of ξ in the plane $z = 0$, that is, $\xi' = (t, \varphi_1, -z_1)$. Thus,

$$\begin{aligned} r &= \{\rho^2 + t^2 - 2\rho t[\cos(\varphi-\varphi_1)] + (z-z_1)^2\}^{\frac{1}{2}} \, , \\ r' &= \{\rho^2 + t^2 - 2\rho t[\cos(\varphi-\varphi_1)] + (z-z_1)^2\}^{\frac{1}{2}} \, . \end{aligned} \tag{21}$$

The differential equations (14) and (19) present us with the integral representation formula by the routine procedure. The required relation is

$$\begin{aligned} w &= v(\rho, z)\cos\varphi \\ &= \frac{1}{4\pi} \int\limits_{z_1=0} \left[-G\frac{\partial v}{\partial z_1}\cos\varphi_1 + v\frac{\partial G}{\partial z_1}\cos\varphi_1 \right]_{z_1=0} dS \, . \end{aligned} \tag{22}$$

Applying the boundary conditions (15), (16), and (19)$_2$ in (22) and using (20), we have

$$\Omega\rho\cos\varphi = \frac{1}{2\pi} \int\limits_0^1 t\phi(t)$$

$$\times \int\limits_0^{2\pi} \frac{\exp ik[\rho^2 + t^2 - 2\rho t\cos(\varphi-\varphi_1)]^{\frac{1}{2}}}{[\rho^2 + t^2 - 2\rho t\cos(\varphi-\varphi_1)]^{\frac{1}{2}}} \cos\varphi_1 \, d\varphi_1 \, dt \, , \tag{23}$$

where

$$\phi(t) = -\frac{\partial v}{\partial z_1}\bigg|_{z_1=0} . \tag{24}$$

Set $\varphi_1 - \varphi = \psi$, proceed as we did in the relations (6.3.2)–(6.3.4), and obtain

$$\Omega\rho = \int_0^1 t\phi(t) K_1(t, \rho) \, dt , \qquad 0 < \rho < 1 , \tag{25}$$

where

$$K_1(t, \rho) = \frac{1}{2\pi} \int_0^{2\pi} \frac{\exp ik(\rho^2 + t^2 - 2\rho t \cos\psi)^{\frac{1}{2}}}{(\rho^2 + t^2 - 2\rho t \cos\psi)^{\frac{1}{2}}} \cos\psi \, d\psi . \tag{26}$$

Finally, we use the identities [21]

$$\frac{\exp ik(\rho^2 + t^2 - 2\rho t \cos\psi)^{\frac{1}{2}}}{(\rho^2 + t^2 - 2\rho \cos\psi)^{\frac{1}{2}}} = \int_0^\infty \frac{p J_0 [p(\rho^2 + t^2 - 2\rho t \cos\psi)^{\frac{1}{2}}]}{\gamma} \, dp, \tag{27}$$

where

$$\gamma = \begin{cases} -i(k^2 - p^2)^{\frac{1}{2}} , & k \geqslant p \\ (p^2 - k^2)^{\frac{1}{2}} , & p \geqslant k \end{cases}$$

and

$$J_0 [p(\rho^2 + t^2 - 2\rho t \cos\psi)^{\frac{1}{2}}] = \sum_{r=0}^\infty (2 - \delta_{0r})(\cos r\psi) J_r(p\rho) J_r(pt) \tag{28}$$

in (26), use the orthogonality of cosine functions, and derive

$$K_1(t, \rho) = \int_0^\infty (p/\gamma) J_1(p\rho) J_1(pt) \, dp . \tag{29}$$

When $\omega \to 0$, we get the corresponding integral equation for the steady rotation of the elastic half-space, and (29) becomes

$$K_1(t, \rho) = \int_0^\infty J_1(p\rho) J_1(pt) \, dp . \tag{30}$$

For the case of an annular disk with inner radius b and outer radius a, the integral equation that corresponds to (25) is of the form

$$\Omega\rho = \int_b^a t\phi(t) K_1(t,\rho)\, dt \,, \qquad b < \rho < a \,. \tag{31}$$

Example 3. *Steady Stokes flow in an unbounded medium.* The Stokes flow equations

$$\nabla^2 \mathbf{q} = \nabla p \,, \qquad \nabla \cdot \mathbf{q} = 0 \tag{32}$$

govern the slow, steady flow of incompressible viscous fluids. These equations have been made dimensionless with the help of the free-stream velocity u and a characteristic length a inherent in the problem. The quantities $\mathbf{q}$ and p stand for the velocity vector and pressure, respectively.

Let S be the surface of a solid B moving in the fluid; then, the boundary conditions are

$$\mathbf{q}(x) = \mathbf{e}_1 \,, \qquad \mathbf{x} \in S \,, \qquad \mathbf{q}(\mathbf{x}) = 0 \qquad \text{as} \quad \mathbf{x} \to \infty \,, \tag{33}$$

where $\mathbf{e}_1$ is the direction of motion of B, taken to be in the x_1 direction. The boundary value problem (32)–(33) can be converted into a Fredholm integral equation of the first kind by defining the Green's tensor $\mathbf{T}_1$ (or T_{1ik}) and Green's vector $\mathbf{p}_1$ (or p_{1j}), which satisfy the mathematical system

$$\nabla^2 \mathbf{T}_1 - \nabla \mathbf{p}_1 = -\mathbf{I}\delta(\mathbf{x}-\boldsymbol{\xi}) \,, \tag{34}$$

$$\nabla \cdot \mathbf{T}_1 = 0 \,, \qquad \mathbf{T}_1 \to 0 \qquad \text{as} \quad \mathbf{x} \to \infty \,, \tag{35}$$

where $\mathbf{I} = \delta_{ij}$, the Kronecker delta.

It follows by direct verification that the system (34)–(35) has the representation formulas

$$\mathbf{T}_1 = (1/8\pi)[\mathbf{I}\nabla^2 \phi - \operatorname{grad}\operatorname{grad}\phi] \,, \qquad \mathbf{p}_1 = -(1/8\pi)\operatorname{grad}\nabla^2 \phi \,, \tag{36}$$

$$\nabla^4 \phi = \nabla^2 \nabla^2 \phi = -8\pi\delta(\mathbf{x}-\boldsymbol{\xi}) \,. \tag{37}$$

The appropriate solution of the biharmonic equation (37) is $\phi = r = |\mathbf{x}-\boldsymbol{\xi}|$. Thus,

$$\mathbf{T}_1 = (1/8\pi)[\mathbf{I}\nabla^2 |\mathbf{x}-\boldsymbol{\xi}| - \operatorname{grad}\operatorname{grad}|\mathbf{x}-\boldsymbol{\xi}|] \,, \tag{38}$$

$$\mathbf{p}_1 = -(1/8\pi)\operatorname{grad}\nabla^2 |\mathbf{x}-\boldsymbol{\xi}| \,. \tag{39}$$

The required integral equation formula now follows by taking the scalar product of (32) by $\mathbf{T}_1$ and of (34) by $\mathbf{q}$ and using the usual steps of subtracting and integrating. In the integral so obtained, there occur terms $\mathbf{T} \cdot \nabla p$ and $\mathbf{q} \cdot \nabla p$ and they can be processed by using the identities

$$\nabla \cdot (\mathbf{q} \mathbf{p}_1) = \mathbf{q} \cdot \nabla \mathbf{p}_1 , \qquad \nabla \cdot (p \mathbf{T}_1) = \mathbf{T}_1 \cdot \nabla p , \tag{40}$$

where we have used the results $\nabla \cdot \mathbf{q} = \nabla \cdot \mathbf{T}_1 = 0$. The final result is

$$\mathbf{q}(\mathbf{x}) = - \int_S \left[\left(\frac{\partial \mathbf{q}}{\partial n} - p\mathbf{n} \right) \cdot \mathbf{T}_1 - \mathbf{q} \cdot \left(\frac{\partial \mathbf{T}_1}{\partial n} - \mathbf{n} \mathbf{p}_1 \right) \right] dS . \tag{41}$$

From equation (34) and the divergence theorem, it follows that, if $\mathbf{q}$ is constant on S, then equation (41) reduces to

$$\mathbf{q}(\mathbf{x}) = - \int_S \mathbf{f} \cdot \mathbf{T}_1 \, dS , \tag{42}$$

where

$$\mathbf{f} = \frac{\partial \mathbf{q}}{\partial n} - p\mathbf{n} . \tag{43}$$

Finally, using the boundary condition $(33)_1$, we have the required integral equation

$$\mathbf{e}_1 = - \int_S \mathbf{f} \cdot \mathbf{T}_1 \, dS . \tag{44}$$

Example 4.   *Steady Oseen flow.*   The dimensionless Oseen equations are

$$\mathscr{R} \, \partial \mathbf{q} / \partial x_1 = - \nabla p + \nabla^2 \mathbf{q} , \qquad \nabla \cdot \mathbf{q} = 0 , \tag{45}$$

$$\mathbf{q} = \mathbf{e}_1 \quad \text{on} \quad S ; \quad \mathbf{q} \to 0 \quad \text{as} \quad \mathbf{x} \to \infty , \tag{46}$$

where $\mathscr{R} = ua/v$ is the Reynolds number, v is the coefficient of kinematic viscosity, and a and u are the same quantities as defined in the previous example. The Green's tensor $\mathbf{T}$ and Green's vector $\mathbf{p}$ for this problem satisfy the system

$$\mathscr{R} \, \partial \mathbf{T} / \partial x_1 = - \nabla \mathbf{p} + \nabla^2 \mathbf{T} + \mathbf{I} \delta (\mathbf{x} - \xi) , \tag{47}$$

$$\nabla \cdot \mathbf{T} = 0 , \qquad \mathbf{T} \to 0 \quad \text{as} \quad \mathbf{x} \to \infty . \tag{48}$$

The corresponding representation formulas are

$$\mathbf{T} = (1/8\pi) [\mathbf{I} \nabla^2 \phi - \operatorname{grad} \operatorname{grad} \phi] , \tag{49}$$

$$\mathbf{p} = -(1/8\pi)\operatorname{grad}[\nabla^2\phi - \mathscr{R}\,\partial\phi/\partial x_1] \,. \tag{50}$$

$$\nabla^2(\nabla^2 - \mathscr{R}\,\partial/\partial x_1)\phi = -8\pi\,\delta(\mathbf{x}-\boldsymbol{\xi}) \,. \tag{51}$$

To solve equation (51) for ϕ, we first use the formula

$$\nabla^2(1/|\mathbf{x}-\boldsymbol{\xi}|) = -4\pi\,\delta(\mathbf{x}-\boldsymbol{\xi}) \tag{52}$$

in it and obtain

$$\nabla^2(\nabla^2 - \mathscr{R}\,\partial/\partial x_1)\phi = 2\nabla^2(1/|\mathbf{x}-\boldsymbol{\xi}|) \,. \tag{53}$$

Thus, if ϕ satisfies

$$\nabla^2\phi - \mathscr{R}\,\partial\phi/\partial x_1 = 2/|\mathbf{x}-\boldsymbol{\xi}| \,, \tag{54}$$

then (51) is satisfied.

On the other hand, we can also write (51) as

$$(\nabla^2 - \mathscr{R}\,\partial/\partial x_1)\nabla^2\phi = -8\pi\,\delta(\mathbf{x}-\boldsymbol{\xi}) \,. \tag{55}$$

By setting

$$\nabla^2\phi = \psi \tag{56}$$

and using the identity

$$(\nabla^2 - \sigma^2)[e^{-\sigma(x_1-\xi_1)}\psi] = e^{-\sigma(x_1-\xi_1)}[\nabla^2 - 2\sigma\,\partial/\partial x_1]\psi \,, \tag{57}$$

we can write (55) as

$$(\nabla^2 - \sigma^2)[e^{-\sigma(x_1-\xi_1)}\psi] = -8\pi e^{-\sigma(x_1-\xi_1)}\delta(\mathbf{x}-\boldsymbol{\xi}) \tag{58}$$

with $\sigma = 2\mathscr{R}$. Now, observe that, by the nature of the Dirac delta function, the factor $e^{-\sigma(x_1-\xi_1)}$ influences the equation only at $x_1 = \xi_1$, where its value is unity. Thus, (58) yields

$$\psi(\mathbf{x}, \boldsymbol{\xi}) = \nabla^2\phi = \frac{2}{|\mathbf{x}-\boldsymbol{\xi}|}\exp\left\{-|\sigma|\left[|\mathbf{x}-\boldsymbol{\xi}| - \frac{\sigma}{|\sigma|}(x_1-\xi_1)\right]\right\}. \tag{59}$$

From (54) and (59), we obtain

$$\frac{\partial\phi}{\partial x_1} = -\frac{1}{\sigma|\mathbf{x}-\boldsymbol{\xi}|}\left(1 - \exp\left\{-|\sigma|\left[|\mathbf{x}-\boldsymbol{\xi}| - \frac{\sigma}{|\sigma|}(x_1-\xi_1)\right]\right\}\right). \tag{60}$$

If we set

$$s = |\mathbf{x}-\boldsymbol{\xi}| - (\sigma/|\sigma|)(x_1-\xi_1) \,, \tag{61}$$

we have

$$\frac{\partial s}{\partial x_1} = \frac{(x_1 - \xi_1)}{|\mathbf{x} - \boldsymbol{\xi}|} - \frac{\sigma}{|\sigma|} = \frac{(x_1 - \xi_1)}{|\mathbf{x} - \boldsymbol{\xi}|} - \frac{|\sigma|}{\sigma},$$

$$= -\frac{|\sigma|}{\sigma|\mathbf{x} - \boldsymbol{\xi}|}\left(|\mathbf{x} - \boldsymbol{\xi}| - \frac{\sigma}{|\sigma|}(x_1 - \xi_1)\right) = -\frac{|\sigma| s}{\sigma|\mathbf{x} - \boldsymbol{\xi}|}. \tag{62}$$

Combining (60)–(62) with

$$\frac{\partial \phi}{\partial x_1} = \frac{\partial \phi}{\partial s}\frac{\partial s}{\partial x_1},$$

we find that

$$d\phi/ds = (1 - e^{-|\sigma|s})/|\sigma| s$$

or

$$\phi = (1/|\sigma|) \int_0^{|\sigma| s} [(1 - e^{-t})/t]\, dt . \tag{63}$$

Thereby, the Green's tensor $\mathbf{T}$ and Green's vector $\mathbf{p}$ are determined.

The integral equation equivalent to the boundary value problem (45)–(46) is now found precisely the way we found the integral equation (41). Indeed, the present formula is

$$\mathbf{q}(\mathbf{x}) = -\int_S \left[\mathbf{T}\cdot\left(\frac{\partial \mathbf{q}}{\partial n} - p\mathbf{n}\right) - \left(\frac{\partial \mathbf{T}}{\partial n} - p\mathbf{n}\right)\cdot\mathbf{q} - \mathscr{R}(\mathbf{T}\cdot\mathbf{q})n_1\right] dS , \tag{64}$$

where n_1 is the x_1 component of the outward normal. When we use the boundary condition $\mathbf{q} = \mathbf{e}_1$ on S, equation (47), and the divergence theorem, we obtain

$$\mathbf{e}_1 = -\int_S \mathbf{T}\cdot\mathbf{f}\, dS , \tag{65}$$

where $\mathbf{f}$ is defined by (43).

Example 5. *Heat conduction.* The boundary value problem

$$\nabla^2 u - k^2 u = -\rho(\mathbf{x}) , \qquad \mathbf{x} \in R_e , \qquad \left.\frac{\partial u}{\partial n}\right|_S = 0 \tag{66}$$

embodies the solution of the heat conduction problem of an infinite expanse of material containing a cavity S on which the temperature gradient is zero. Our aim is to give the system (66) an integral-equation formulation. For this purpose, we assume that u can be represented as the sum of a volume potential and a single-layer potential

$$4\pi u(\mathbf{x}) = \int_{R_e} \rho(\xi) E(\mathbf{x};\xi)\, dV + \int_S \sigma(\xi) E(\mathbf{x};\xi)\, dS\,, \tag{67}$$

where $E(\mathbf{x};\xi) = [\exp(-k|\mathbf{x}-\xi|)]/|\mathbf{x}-\xi|$ and σ is an unknown source density.

The next step is to take the normal derivative of both sides of equation (67), let $\mathbf{x}$ approach S, and use relation (6.6.8). The result is

$$0 = \int_{R_e} \rho(\xi)\frac{\partial E(\mathbf{x};\xi)}{\partial n}\, dV - 2\pi\sigma(\mathbf{x}) + \int_S \sigma(\xi)\frac{\partial E(\mathbf{x};\xi)}{\partial n}\, dS\,, \tag{68}$$

which is a Fredholm integral equation of the second kind in $\sigma(\mathbf{x})$.

The reader who is familiar with the theory of heat conduction is advised to formulate the corresponding problem of the composite medium into a Fredholm integral equation of the first kind.

EXERCISES

1. Show that when S is the surface of a unit sphere (a) the solution of the integral equation

$$\rho z \cos\varphi = \int_S \sigma(\rho_1, z_1)|\mathbf{x}-\xi|^{-1}\cos\varphi_1\, dS$$

is $\sigma = (5/12)P_2{}^1(\cos\theta)$; (b) the solution of the integral equation

$$(1/2)\rho z^2 \cos\varphi = (3/8\pi)\int_S P_1{}^1(\cos\theta_1)|\mathbf{x}-\xi|\cos\varphi_1\, dS$$

$$+ \int_S \sigma(\rho_1, z_1)|\mathbf{x}-\xi|^{-1}\cos\varphi_1\, dS$$

is $\sigma = (1/4\pi)[(3/2)P_1{}^1(\cos\theta)+(7/15)P_3{}^1(\cos\theta)]$.

In the above relations, (ρ, φ, z) are cylindrical polar coordinates.

Hints: (i) Use the formula

$$|\mathbf{x}-\xi| = r_> \sum_{v=1}^{\infty} \left(\frac{x^2}{2v+3} - \frac{1}{2v-1} \right) x^v P_v(\cos \gamma) ,$$

where

$$x = r_</r_> , \qquad r_< = \min(r, r_1) , \qquad r_> = \max(r, r_1) ,$$

$$\cos \gamma = \cos \theta \cos \theta_1 + \sin \theta \sin \theta_1 \cos(\varphi - \varphi_1) ,$$

while (r, φ, θ) and $(r_1, \varphi_1, \theta_1)$ are the spherical polar coordinates of $\mathbf{x}$ and ξ, respectively.

(ii) On the surface of the unit sphere, $\rho = \sin \theta$, $z = \cos \theta$.

2. Show that the solutions of the integral equations

$$\rho \cos \varphi = \int_S \sigma(\rho_1, z_1) |\mathbf{x}-\xi|^{-1} \cos \varphi_1 \, dS$$

and

$$0 = \int_S \sigma'(\rho_1, z_1) |\mathbf{x}-\xi|^{-1} \cos \varphi_1 \, dS$$

$$+ \tfrac{1}{2} \int_S \sigma(\rho_1, z_1) |\mathbf{x}-\xi| \cos \varphi_1 \, dS ,$$

where S is the surface of a thin circular disk of unit radius, are

$$\sigma = 2\rho/\pi^2 (1-\rho^2)^{1/2} , \qquad \sigma' = \rho(2-\rho^2)/3\pi^2 (1-\rho^2)^{1/2} .$$

3. Starting with the Cauchy integral formula for an analytic function

$$f(z) = (1/2\pi i) \int_C [f(t)/(t-z)] \, dt ,$$

where C is the circumference $|z| = a$ and z is in the interior of C, and using the formula

$$0 = (1/2\pi i) \int_C [f(t)/(t-z^*)] \, dt , \qquad z^* = a^2/\bar{z} ,$$

which is a result of Cauchy theorem because the image part z^* (of z) is exterior to C, derive the Poisson integral formula in a plane:

$$u(\rho, \varphi) = \frac{a^2 - \rho^2}{2\pi} \int_0^{2\pi} \frac{f(a, \theta) \, d\theta}{a^2 - 2a\rho [\cos(\theta - \varphi)] + a^2} .$$

4. Discuss the single-layer and double-layer potentials in two-dimensional potential theory by starting with the formula

$$E(\mathbf{x};\boldsymbol{\xi}) = (1/2\pi)\log(1/|\mathbf{x}-\boldsymbol{\xi}|)$$

instead of (6.2.2).

5. With the help of the results obtained in the previous exercise, prove that the solution of the interior Dirichlet problem in two dimensions can be written as

$$w(\mathbf{x}) = \int_C [(\cos\psi)/|\mathbf{x}-\boldsymbol{\xi}|]\,\tau(\boldsymbol{\xi})\,d\ell\;,$$

where ψ is the angle between $(\mathbf{x}-\boldsymbol{\xi})$ and $\mathbf{n}$; and $d\ell$ denotes the element of the arc length along the curve C.

6. Proceeding as in Example 1 of Section 6.7, show that the following boundary value problem,

$$(\nabla^2 + k^2)u_s = 0\;,$$

$$\partial u_s/\partial z = -\partial u_i/\partial z \quad \text{on} \quad z = 0\;, \quad 0 \leqslant \rho \leqslant a\;,$$

u_s and $\partial u_s/\partial z$ are continuous across $z = 0$, $\quad a < \rho < \infty$,

and u_s satisfies the radiation condition, reduces to the integral equation

$$\frac{\partial u_i}{\partial z}(\rho,\varphi,0) = \int_0^a t\sigma(t)\left[\frac{\partial^2}{\partial z^2}\left(\frac{\exp ik\,|\mathbf{x}-\boldsymbol{\xi}|}{|\mathbf{x}-\boldsymbol{\xi}|}\right)\right]_{z_1=z=0} d\varphi_1\,dt\;,$$

where $\sigma(t) = (1/4\pi)(u_s|_{z_1=0+} - u_s|_{z=0-})$. This boundary value problem embodies the solution of the diffraction of a plane wave by a perfectly rigid, circular thin disk.

7. By following the method and notation of Example 3 in Section 6.7, prove that the integral representation formula for the velocity vector when the fluid is bounded by a vessel Σ is

$$\mathbf{q}(\mathbf{x}) = -\int\left[\left(\frac{\partial\mathbf{q}}{\partial n} - p\mathbf{n}\right)\cdot\mathbf{T} - \mathbf{q}\cdot\left(\frac{\partial\mathbf{T}}{\partial n} - \mathbf{n}p\right)\right]dS\;.$$

Substituting the boundary condition $\mathbf{q} = \mathbf{e}_1$ on S gives the Fredholm integral equation

$$\mathbf{e}_1 = -\int_S \mathbf{f} \cdot \mathbf{T} \, dS \,.$$

Here, $\mathbf{T}$ satisfies the boundary value problem

$$\nabla^2 \mathbf{T} - \operatorname{grad} p = \mathbf{I}\delta(\mathbf{x}-\boldsymbol{\xi}) \,, \qquad \nabla\cdot\mathbf{T} = 0 \,, \qquad \mathbf{T} = 0 \quad \text{on} \quad \Sigma \,.$$

8. The dimensionless equations of elastostatics are

$$(\lambda+\mu)\operatorname{grad}\theta + \mu\nabla^2\mathbf{u} = 0 \,, \qquad \theta = \operatorname{div}\mathbf{u} \,,$$

where $\mathbf{u}$ $(u_i.\ i=1,2,3)$ is the displacement vector, and λ and μ are Lamé constants of the medium. The above equations have been made dimensionless by introducing a characteristic geometric length a.

Consider the uniform translation of a light, rigid body B with surface S_1, which is embedded in an infinite homogeneous and isotropic medium. Prove that the corresponding Green's tensor $\mathbf{T}_1$ and Green's dilation vector $\boldsymbol{\theta}_1$ are given by the formulas

$$T_{1ij} = \frac{1}{8\pi}\left[\frac{\lambda+3\mu}{\lambda+2\mu}\frac{\delta_{ij}}{|\mathbf{x}-\boldsymbol{\xi}|} + \frac{\lambda+\mu}{\lambda+2\mu}\frac{(x_i-\xi_i)(x_j-\xi_j)}{|\mathbf{x}-\boldsymbol{\xi}|^3}\right] \,,$$

$$\theta_{1i} = -\frac{1}{4\pi}\frac{\mu}{\lambda+2\mu}\left[\frac{x_i-\xi_i}{|\mathbf{x}-\boldsymbol{\xi}|^3}\right] \,.$$

Also prove that the integral representation formula is

$$\mathbf{u}(p) = -\int_S \left\{\left[\mu\frac{d\mathbf{u}}{\partial n} + (\lambda+\mu)\theta\mathbf{n}\right]\cdot\mathbf{T}\right.$$

$$\left. -\mathbf{u}\cdot\left[\mu\frac{d\mathbf{T}_1}{dn} + (\lambda+\mu)\boldsymbol{\theta}_1\,\mathbf{n}\right]\right\}\,dS \,.$$

Using the boundary conditions

$$\mathbf{u} = (d_0/a)\,\mathbf{e} \qquad \text{on} \quad S \,; \qquad \mathbf{u} \to 0 \quad \text{at} \quad \infty \,;$$

where d_0 is the magnitude of the translation and $\mathbf{e}$ is the unit direction along the translation of B, show that the Fredholm integral equation that embodies the solution of the above problem is

$$(d_0/a)\,\mathbf{e} = -\int \mathbf{f}\cdot\mathbf{T}_1 \, dS \,,$$

where

$$\mathbf{f} = \mu(d\mathbf{u}/dn) + (\lambda+\mu)\theta\mathbf{n} \,.$$

Extend the above results to a bounded medium.

9. Proceed as in Example 1 of Section 6.3 and Example 1 of Section 6.7 and give the integral-equation formulation of these problems for a spherical cap.

10. The Schrödinger equation

$$\nabla^2 \psi(\mathbf{x}) - (2m/\hbar^2)V(\mathbf{x})\psi(\mathbf{x}) + k^2 \psi(\mathbf{x}) = 0$$

with boundary condition that $\psi(\mathbf{x})\exp(-iEt/\hbar)$ represents an incident plane wave, with wave vector $\mathbf{k}_0$, as well as the condition that we have an outgoing wave as $\mathbf{x} \to \infty$, describes the quantum mechanical theory of scattering by a potential $V(\mathbf{x})$. Here, $k^2 = k_0{}^2 = 2mE/\hbar^2$, and other quantities have their usual meaning.

Use the method of Section 6.6 and prove that we can transform this scattering problem into the integral equation

$$\psi(\mathbf{x}) = (\exp i\mathbf{k}_0 \cdot \mathbf{x}) - \frac{m}{2\pi\hbar^2} \int_R \frac{\exp ik|\mathbf{x}-\boldsymbol{\xi}|}{|\mathbf{x}-\boldsymbol{\xi}|} V(\boldsymbol{\xi})\psi(\boldsymbol{\xi})\,dV .$$

Find its solution by the method of successive approximations (the approximation obtained in this way is called the Born approximation).

7.1. INTRODUCTION

A kernel $K(s, t)$ is symmetric (or complex symmetric or Hermitian) if

$$K(s, t) = K^*(t, s) , \tag{1}$$

where the asterisk denotes the complex conjugate. In the case of a real kernel, the symmetry reduces to the equality

$$K(s, t) = K(t, s) . \tag{2}$$

We have seen in the previous two chapters that the integral equations with symmetric kernels are of frequent occurrence in the formulation of physically motivated problems.

We claim that if a kernel is symmetric then all its iterated kernels are also symmetric. Indeed,

$$K_2(s, t) = \int K(s, x) K(x, t) \, dx = \int K^*(t, x) K^*(x, s) \, dx = K_2^*(t, s) .$$

Again, if $K_n(s, t)$ is symmetric, then the recursion relation gives

$$K_{n+1}(s, t) = \int K(s, x) K_n(x, t) \, dx$$

$$= \int K_n^*(t, x) K^*(x, s) \, dx = K_{n+1}^*(t, s) . \tag{3}$$

The proof of our claim follows by induction. Note that the *trace* $K(s,s)$ of a symmetric kernel is always real because $K(s,s) = K^*(s,s)$. Similarly, the traces of all iterates are also real.

Complex Hilbert Space

We present here a brief review of the properties of the complex Hilbert space $\mathscr{L}_2(a,b)$, which shall be needed in the sequel. This discussion is valid for real $\mathscr{L}_2$-space as a special case. A linear space of infinite dimension with inner product (or scalar product) (x,y) which is a complex number satisfying (a) the definiteness axiom $(x,x) > 0$ for $x \neq 0$; (b) the linearity axiom $(\alpha x_1 + \beta x_2, y) = \alpha(x_1, y) + \beta(x_2, y)$, where α and β are arbitrary complex numbers; and (c) the axiom of (Hermitian) symmetry $(y,x) = (x,y)^*$; is called a complex Hilbert space.

Let H be the set of complex-valued functions $\phi(t)$ defined in the interval (a,b) such that

$$\int |\phi(t)|^2 \, dt < \infty . \tag{4}$$

Furthermore, let us define the inner product by the bilinear form (1.6.1):

$$(\phi, \psi) = \int \phi(t) \psi^*(t) \, dt . \tag{5}$$

Then, H is a linear and complex Hilbert space $\mathscr{L}_2(a,b)$ (or $\mathscr{L}_2$). The norm $\|\phi\|$ as defined by (1.6.2),

$$\|\phi\| = \left(\int |\phi(t)|^2 \, dt \right)^{\frac{1}{2}} \tag{6}$$

is called the norm that generates the natural metric

$$d(\phi, \psi) = \|\phi - \psi\| = (\phi - \psi, \phi - \psi)^{\frac{1}{2}} . \tag{7}$$

The Schwarz and Minkowskii inequalities as given by (1.6.3) and (1.6.4) are

$$|(\phi, \psi)| \leq \|\phi\| \, \|\psi\| , \tag{8}$$

$$\|\phi + \psi\| \leq \|\phi\| + \|\psi\| . \tag{9}$$

Also recall the definition of an $\mathscr{L}_2$-kernel as given by (1.2.2)–(1.2.4).

Another concept that is fundamental in the theory of Hilbert spaces is the concept of completeness. A metric space is called complete if every Cauchy sequence of functions in this space is a convergent sequence. If a metric space is not complete, then there is a simple way to add elements to this space to make it complete. A Hilbert space is an inner-product linear space that is complete in its natural metric. The completeness of $\mathscr{L}_2$-spaces plays an important role in the theory of linear operators such as the Fredholm operator K,

$$K\phi = \int K(s,t)\,\phi(t)\,dt \ . \tag{10}$$

The operator adjoint to K is

$$K^*\psi = \int K^*(t,s)\,\psi(t)\,dt \ . \tag{11}$$

The operators (10) and (11) are connected by the interesting relationship

$$(K\phi,\psi) = (\phi, K^*\psi) \ , \tag{12}$$

which is proved as follows:

$$
\begin{aligned}
(K\phi,\psi) &= \int \psi^*(s)\left[\int K(s,t)\,\phi(t)\,dt\right] ds \\
&= \int \phi(t)\left[\int K(s,t)\,\psi^*(s)\,ds\right] dt \\
&= \int \phi(s)\left[\int K(t,s)\,\psi^*(t)\,dt\right] ds \\
&= \int \phi(s)\left[\int K^*(t,s)\,\psi(t)\,dt\right]^* ds \\
&= (\phi, K^*\psi) \ .
\end{aligned}
$$

For a symmetric kernel, this result becomes

$$(K\phi,\psi) = (\phi, K\psi) \ , \tag{13}$$

i.e., a symmetric operator is self-adjoint. Note that permutation of factors in a scalar product is equivalent to taking the complex conjugate, i.e., $(\phi, K\phi) = (K\phi, \phi)^*$. Combining this with (13), we find that, *for a symmetric kernel, the inner product $(K\phi, \phi)$ is always real*; the converse is also true and forms Exercise 1 at the end of this chapter.

An Orthonormal System of Functions

Systems of orthogonal functions play an important role in the theory of integral equations and their applications. A finite or an infinite set $\{\phi_k\}$ is said to be an orthogonal set if $(\phi_i, \phi_j) = 0$, $i \neq j$. If none of the elements of this set is a zero vector, then it is said to be a proper orthogonal set. A set is orthonormal if

$$(\phi_i, \phi_j) = \begin{cases} 0, & i \neq j, \\ 1, & i = j. \end{cases}$$

Individual functions ϕ for which $\|\phi\| = 1$ are said to be normalized.

Given a finite or an infinite (denumerable) independent set of functions $\{\psi_1, \psi_2, ..., \psi_k, ...\}$, we can construct an orthonormal set $\{\phi_1, \phi_2, ..., \phi_k, ...\}$ by the well-known Gram–Schmidt procedure as follows.

Let $\phi_1 = \psi_1 / \|\psi_1\|$. To obtain ϕ_2, we define

$$w_2(s) = \psi_2(s) - (\psi_2, \phi_1)\phi_1 .$$

The function w_2 is clearly orthogonal to ϕ_1; thus, ϕ_2 is obtained by setting $\phi_2 = w_2 / \|w_2\|$. Continuing this process, we have

$$w_k(s) = \psi_k(s) - \sum_{i=1}^{k-1} (\psi_k, \phi_i)\phi_i , \qquad \phi_k = w_k / \|w_k\| .$$

We have, thereby, obtained an equally numerous set of orthonormal functions. In case we are given a set of orthogonal functions, we can convert it into an orthonormal set simply by dividing each function by its norm.

Starting from an arbitrary orthonormal system, it is possible to construct the theory of Fourier series, analogous to the theory of trigonometric series. Suppose we want to find the best approximation of an arbitrary function $\psi(x)$ in terms of a linear combination of an orthonormal set $(\phi_1, \phi_2, ..., \phi_n)$. By the best approximation, we mean that we choose the coefficients $\alpha_1, \alpha_2, ..., \alpha_n$ such as to minimize $\|\psi - \sum_{i=1}^{n} \alpha_i \phi_i\|$, or, what is equivalent, to minimize $\|\psi - \sum_{i=1}^{n} \alpha_i \phi_i\|^2$.

Now, for any $\alpha_1, ..., \alpha_n$, we have

$$\left\| \psi - \sum_{i=1}^{n} \alpha_i \phi_i \right\|^2 = \|\psi\|^2 + \sum_{i=1}^{n} |(\psi, \phi_i) - \alpha_i|^2 - \sum_{i=1}^{n} |(\psi, \phi_i)|^2 . \quad (14)$$

It is obvious that the minimum is achieved by choosing $\alpha_i = (\psi, \phi_i) = a_i$ (say). The numbers a_i are called the Fourier coefficients of the function $\psi(s)$ relative to the orthonormal system $\{\phi_i\}$. In that case, the relation (14) reduces to

$$\left\| \psi - \sum_{i=1}^{n} \alpha_i \phi_i \right\|^2 = \|\psi\|^2 - \sum_{i=1}^{n} |a_i|^2 . \quad (15)$$

Since the left side is nonnegative, we have

$$\sum_{i=1}^{n} |a_i|^2 \leqslant \|\psi\|^2 , \quad (16)$$

which, for the infinite set $\{\phi_i\}$, leads to the *Bessel inequality*

$$\sum_{i=1}^{\infty} |a_i|^2 \leqslant \|\psi\|^2 . \quad (17)$$

Suppose we are given an infinite orthonormal system $\{\phi_i(s)\}$ in $\mathscr{L}_2$, and a sequence of constants $\{\alpha_i\}$, then the convergence of the series $\sum_{k=1}^{\infty} |\alpha_k^2|$ is evidently a necessary condition for the existence of an $\mathscr{L}_2$-function $f(s)$ whose Fourier coefficients with respect to the system ϕ_i are α_i. It so happens that this condition is also sufficient and the result is embodied in the Riesz–Fischer theorem, which we state as follows without proof.

Riesz–Fischer Theorem.   If $\{\phi_i(s)\}$ is a given orthonormal system of functions in $\mathscr{L}_2$ and $\{\alpha_i\}$ is a given sequence of complex numbers such that the series $\sum_{k=1}^{\infty} |\alpha_k^2|$ converges, then there exists a unique function $f(s)$ for which α_i are the Fourier coefficients with respect to the orthonormal system $\{\phi_i\}$ and to which the Fourier series converges in the mean, that is,

$$\left\| f(s) - \sum_{i=1}^{n} \alpha_i \phi_i \right\| \to 0 \qquad \text{as} \quad n \to \infty .$$

If an orthonormal system of functions ϕ_i can be found in $\mathscr{L}_2$-space such that every other element of this space can be represented linearly in terms of this system, then it is called an orthonormal basis. The concepts of an orthonormal basis and a complete system of orthonormal functions are equivalent. Indeed, if any of the following criteria are met, the orthonormal set $\{\phi_1, ..., \phi_k, ...\}$ is complete.

(a)   For every function ψ in $\mathcal{L}_2$,

$$\psi = \sum_i (\psi, \phi_i) \phi_i = \sum_i a_i \phi_i . \tag{18}$$

(b)   For every function ψ in $\mathcal{L}_2$,

$$\|\psi\|^2 = \sum_{i=1}^{\infty} |(\psi, \phi_i)|^2 . \tag{19}$$

This is called *Parseval's identity*.

(c)   The only function ψ in $\mathcal{L}_2$ for which all the Fourier coefficients vanish is the trivial function (zero function).

(d)   There exists no function ψ in $\mathcal{L}_2$ such that $\{\psi, \phi_1, ..., \phi_k, ...\}$ is an orthonormal set.

The equivalence of these different criteria can be easily proved.

One frequently encounters Fourier series of somewhat more general character. Let $r(t)$ be a continuous, real, and nonnegative function in the interval (a, b). We say that the set of functions $\{\phi_i\}$ is orthonormal with weight $r(t)$ if

$$\int r(t) \phi_j(t) \phi_k(t) \, dt = \begin{cases} 0, & j \neq k, \\ 1, & j = k. \end{cases} \tag{20}$$

The Fourier expansions in terms of such functions are treated by introducing a new inner product

$$(\phi, \psi) = \int r(t) \phi(t) \psi^*(t) \, dt \tag{21}$$

with the corresponding norm

$$\|\phi\|_r = \left[\int r(t) \phi(t) \phi^*(t) \, dt \right]^{1/2} . \tag{22}$$

The space of functions for which $\|\phi\|_r < \infty$ is a Hilbert space and all the above results hold.

Some examples of the complete orthogonal and orthonormal systems are listed below.

(a)   The system $\phi_k(s) = (2\pi)^{-1/2} e^{iks}$ is orthonormal, where k is any integer $-\infty < k < \infty$.

(b)   The Legendre polynomials

$$P_0(s) = 1, \qquad P_n(s) = \frac{1}{2^n n!} \frac{d^n (s^2 - 1)^n}{ds^n}, \qquad n = 1, 2, \ldots$$

are orthogonal in the interval $(-1, 1)$. Indeed,

$$\int_{-1}^{1} P_j(s) P_k(s) \, ds = \begin{cases} 0, & j \neq k, \\ 2/(2k+1), & j = k. \end{cases} \tag{23}$$

(c)   Let $\alpha_{k,n}$ denote the positive zeros of the Bessel function $J_n(s)$, $k = 1, 2, \ldots$; $n > -1$. The system of functions $J_n(\alpha_{k,n} s)$ is orthogonal with weight $r(s) = s$ in the interval $(0, 1)$:

$$\int_{0}^{1} s J_n(\alpha_{j,n} s) J_n(\alpha_{k,n} s) \, ds = \begin{cases} 0, & j \neq k, \\ J_{n+1}^{2}(\alpha_{k,n}), & j = k. \end{cases} \tag{24}$$

A Complete Two-Dimensional Orthonormal Set over the Rectangle $a \leqslant s \leqslant b, c \leqslant t \leqslant d$

Let $\{\phi_i(s)\}$ be a complete orthonormal set over $a \leqslant s \leqslant b$, and let $\{\psi_i(t)\}$ be a complete orthonormal set over $c \leqslant t \leqslant d$. Then, we claim that the set $\phi_1(s)\psi_1(t)$, $\phi_1(s)\psi_2(t), \ldots, \phi_2(s)\psi_1(t)$, .. is a complete two-dimensional orthonormal set over the rectangle $a \leqslant s \leqslant b, c \leqslant t \leqslant d$.

The fact that the sequence of two-dimensional functions $\{\phi_i(s)\psi_j(t)\}$ is an orthonormal set follows readily by integrating over the rectangle. The completeness is proved by showing that every continuous function $F(s, t)$ with finite norm $\|F\|$, whose Fourier coefficients with respect to the set $\{\phi_i(s)\psi_j(t)\}$ are all zero vanishes identically over the rectangle.

For details on the results and ideas of this entire section, the reader is referred to [3, 19].

7.2. FUNDAMENTAL PROPERTIES OF EIGENVALUES AND EIGENFUNCTIONS FOR SYMMETRIC KERNELS

We have discussed the eigenvalues and eigenfunctions for integral equations in the previous chapters. In Chapters 2 and 4, we found that the eigenvalues of an integral equation are the zeros of certain determinants. It may happen that no such zeros exist, so that the kernel has no eigenvalues. There are many kernels for which there are no eigenvalues. Consider, for example, the homogeneous equation (cf. Exercise 3, Chapter 2)

$$g(s) = \lambda \int_0^{2\pi} (\sin s \cos t) g(t)\, dt = \lambda(\sin s) \int_0^{2\pi} g(t) \cos t\, dt \, . \tag{1}$$

Its solution must clearly be of the form $g(s) = A \sin s$. Substituting this in (1) yields

$$A \sin s = \lambda(\sin s) \int_0^{2\pi} A(\sin t \cos t)\, dt = 0 \, .$$

Thus, the kernel $K(s, t) = \sin s \cos t$ $(0 \leqslant s \leqslant 2\pi, \ 0 \leqslant t \leqslant 2\pi)$ has no eigenvalues. For a symmetric kernel that is nonnull (i.e., that is not identically zero), the above phenomenon cannot occur. Indeed, a *symmetric kernel possesses at least one eigenvalue.* The proof is briefly discussed in Section 7.8.

An eigenvalue is simple if there is only one corresponding eigenfunction, otherwise the eigenvalues are degenerate. The spectrum of the kernel K is the set of all its eigenvalues. In this terminology, the above assumption states that *the spectrum of a symmetric kernel is never empty.*

The following are some important properties of the symmetric integral equations

$$\lambda \int K(s, t) g(t)\, dt = f(s), \quad \text{or} \quad \lambda Kg = f; \quad K(s, t) = K^*(t, s). \tag{2}$$

1. The eigenvalues of a symmetric kernel are real. Let λ and $\phi(s)$ be an eigenvalue and a corresponding eigenfunction of the kernel $K(s, t)$. This means

$$\phi(s) - \lambda K\phi(s) = 0 \, . \tag{3}$$

Multiply (3) by $\phi^*(s)$ and integrate with respect to s from a to b and derive the relation

$$\|\phi(s)\|^2 - \lambda(K\phi, \phi) = 0$$

or

$$\lambda = \|\phi(s)\|^2/(K\phi, \phi) \, . \tag{4}$$

Since both the numerator and denominator are real, we have the required result.

2. The eigenfunctions of a symmetric kernel, corresponding to different eigenvalues, are orthogonal. If ϕ_1 and ϕ_2 are eigenfunctions corresponding, respectively, to the eigenvalues λ_1 and λ_2, we have

$$\phi_1 - \lambda_1 K\phi_1 = 0, \qquad \phi_2 - \lambda_2 K\phi_2 = 0. \tag{5}$$

Since λ_2 is real, the second equation in (5) may be written as $\phi_2{}^* - \lambda_2 K^* \phi_2{}^* = 0$. Then, by suitable multiplication and integration it follows that

$$\lambda_1 \lambda_2 \left[\iint \phi_2{}^*(s) K(s,t) \phi_1(t)\, dt\, ds \right.$$
$$\left. - \iint \phi_1(s) K^*(s,t) \phi_2{}^*(t)\, dt\, ds \right] = (\lambda_2 - \lambda_1)(\phi_1, \phi_2). \tag{6}$$

But $K(s,t) = K^*(t,s)$, and the left side vanishes identically, and because $\lambda_1 \neq \lambda_2$, the right side becomes $(\phi_1, \phi_2) = 0$.

3. The multiplicity of any nonzero eigenvalue is finite for every symmetric kernel for which $\iint |K(s,t)|^2\, ds\, dt$ is finite. Let the functions $\phi_{1\lambda}(s), \phi_{2\lambda}(s), ..., \phi_{n\lambda}(s) ...$ be the linearly independent eigenfunctions which correspond to a nonzero eigenvalue λ. By the Gram–Schmidt procedure, we can find linear combinations of these functions which form an orthonormal system $\{u_{k\lambda}(s)\}$. Then, the corresponding complex conjugate system $\{u_{k\lambda}{}^*\}$ also forms an orthonormal system. Let

$$K(s,t) \sim \sum_i a_i u_{i\lambda}{}^*(t),$$

$$a_i = \int K(s,t) u_{i\lambda}(t)\, dt = \lambda^{-1} u_{i\lambda}(s)$$

be the series associated with the kernel $K(s,t)$ for a fixed s. Applying Bessel's inequality to this series, we have

$$\int |K(s,t)|^2\, dt \geqslant \sum_i (\lambda^{-1})^2 |u_{i\lambda}(s)|^2,$$

which, when integrated with respect to s, yields

$$\iint |K(s,t)|^2\, ds\, dt \geqslant \sum_i (\lambda^{-1})^2. \tag{7}$$

The right side of (7) is $m(\lambda^{-1})^2$, where m is the multiplicity of λ. Since the left side is finite, it follows that m is finite.

4. The sequence of eigenfunctions of a symmetric kernel can be made orthonormal. Suppose that, corresponding to a certain eigenvalue, there are m linearly independent eigenfunctions. In view of the linearity of the integral operator, every linear combination of these functions is also an eigenfunction. Thus, by the Gram–Schmidt procedure, we can get equally numerous eigenfunctions which are orthonormal.

On the other hand, for different eigenvalues, the corresponding eigenfunctions are orthogonal and can be readily normalized. Combining these two facts, we have the proof of the above property.

5. The eigenvalues of a symmetric $\mathscr{L}_2$-kernel form a finite or an infinite sequence $\{\lambda_n\}$ with no finite limit point. If we include each eigenvalue in the sequence a number of times equal to its multiplicity, then

$$\sum_{n=1}^{\infty} (\lambda_n^{-1})^2 \leqslant \iint |K(s,t)|^2 \, ds \, dt \ . \tag{8}$$

Let $\{u_k(s)\}$ be the orthonormal eigenfunctions corresponding to different (nonzero) eigenvalues λ_i. Then, proceeding as in the proof of the property 3 and applying the Bessel inequality, we have

$$\sum_i (\lambda_i^{-1})^2 \leqslant \iint |K(s,t)|^2 \, ds \, dt < \infty \ .$$

Hence, if there exists an enumerable infinity of λ_i, then we must have $\sum_{n=1}^{\infty} (\lambda_n^{-1})^2 < \infty$. It follows that $\lim(1/\lambda_i) \to 0$ and ∞ is the only limit point of the eigenvalues.

6. The set of eigenvalues of the second iterated kernel coincide with the set of squares of the eigenvalues of the given kernel.

Note that the symmetry of the kernel shall not be assumed to prove this result.

Let λ be an eigenvalue of K with corresponding eigenfunction $\phi(s)$, that is, $(I - \lambda K)\phi = 0$, where I is the identity operator. When we operate on both sides of this equation with the operator $(I + \lambda K)$, we obtain $(I - \lambda^2 K^2)\phi = 0$ or

$$\phi(s) - \lambda^2 \int K_2(s,t)\phi(t) \, dt = 0 \ , \tag{9}$$

which proves that λ^2 is an eigenvalue of the kernel $K_2(s,t)$.

Conversely, let $\mu = \lambda^2$ be an eigenvalue of the kernel $K_2(s,t)$, with $\phi(s)$ as the corresponding eigenfunction. Then, $(I - \lambda^2 K^2)\phi = 0$ or

$$(I - \lambda K)(I + \lambda K)\phi = 0 . \tag{10}$$

If λ is an eigenvalue of K, then the above property is proved. If not, let us set $(I + \lambda K)\phi = \phi'(s)$ in (10) and obtain $(I - \lambda K)\phi'(s) = 0$. Since we have assumed that λ is not an eigenvalue of K, it follows that $\phi'(s) \equiv 0$, or equivalently $(1 + \lambda K)\phi = 0$. Thus, $-\lambda$ is an eigenvalue of the kernel K and the above property is proved.

We can extend the above result to the nth iterate. The set of eigenvalues of the kernel $K_n(s,t)$ coincide with the set of nth powers of the eigenvalues of the kernel $K(s,t)$.

7. If λ_1 is the smallest eigenvalue of the kernel K, then

$$1/|\lambda_1| = \max [|(K\phi, \phi)|/\|\phi\|]$$

$$= \max [|\iint K(s,t)\phi(t)\phi^*(s)\, dt\, ds|/\|\phi\|] \tag{11}$$

or equivalently,

$$1/|\lambda_1| = \max(K\phi, \phi) , \qquad \|\phi\| = 1 . \tag{12}$$

This maximum value is attained when $\phi(s)$ is an eigenfunction of the symmetric $\mathcal{L}_2$-kernel corresponding to the smallest eigenvalue. For proof, see Section 7.8.

7.3. EXPANSION IN EIGENFUNCTIONS AND BILINEAR FORM

We now discuss the results concerning the expansion of a symmetric kernel and of functions represented in a certain sense by the kernel, in terms of its eigenfunctions and the eigenvalues. Recall that we meet a similar situation when we deal with a Hermitian matrix. For instance, if A is a Hermitian matrix, then there is a unitary matrix U such that $U^{-1}AU$ is diagonal. This means that, by transforming to an orthonormal basis of the vector space consisting of the eigenfunctions of A, the matrix representing the operator A becomes diagonal.

Let $K(s,t)$ be a nonnull, symmetric kernel which has a finite or an infinite number of eigenvalues (always real and nonzero). We order them in the sequence

$$\lambda_1, \lambda_2, ..., \lambda_n, ... \tag{1}$$

in such a way that each eigenvalue is repeated as many times as its multiplicity. We further agree to denumerate these eigenvalues in the order that corresponds to their absolute values, i.e.,

$$0 < |\lambda_1| \leqslant |\lambda_2| \leqslant \cdots \leqslant |\lambda_n| \leqslant |\lambda_{n+1}| \leqslant \cdots .$$

Let

$$\phi_1(s), \phi_2(s), ..., \phi_n(s), ... \tag{2}$$

be the sequence of eigenfunctions corresponding to the eigenvalues given by the sequence (1) and arranged in such a way that they are no longer repeated and are linearly independent in each group corresponding to the same eigenvalue. Thus, to each eigenvalue λ_k in (1) there corresponds just one eigenfunction $\phi_k(s)$ in (2). According to the property 4 of the previous section, we assume that they have been orthonormalized.

Now, we have assumed that a symmetric $\mathcal{L}_2$-kernel has at least one eigenvalue, say λ_1. Then $\phi_1(s)$ is the corresponding eigenfunction. It follows that the "truncated" symmetric kernel

$$K^{(2)}(s, t) = K(s, t) - [\phi_1(s)\phi_1{}^*(t)/\lambda_1]$$

is nonnull and it will also have at least one eigenvalue λ_2 (we choose the smallest if there are more) with corresponding normalized eigenfunction $\phi_2(s)$. The function $\phi_1(s) \neq \phi_2(s)$ even if $\lambda_1 = \lambda_2$, since

$$\int K^{(2)}(s, t)\phi_1(t)\, dt = \int K(s, t)\phi_1(t)\, dt$$

$$- \frac{\phi_1(s)}{\lambda_1} \int \phi_1(t)\phi_1{}^*(t)\, dt$$

$$= \frac{\phi_1(s)}{\lambda_1} - \frac{\phi_1(s)}{\lambda_1} = 0 .$$

Similarly, the third truncated kernel

$$K^{(3)}(s, t) = K^{(2)}(s, t) - \frac{\phi_2(s)\phi_2{}^*(t)}{\lambda_2}$$

$$= K(s, t) - \sum_{k=1}^{2} \frac{\phi_k(s)\phi_k{}^*(t)}{\lambda_k}$$

gives the third eigenvalue λ_3 and eigenfunction $\phi_3(s)$. Continuing in this way, we end up with two possibilities: either this process terminates after n steps, that is, $K^{(n+1)}(s, t) \equiv 0$, and the kernel $K(s, t)$ is a degenerate kernel,

$$K(s, t) = \sum_{k=1}^{n} \frac{\phi_k(s) \phi_k^*(t)}{\lambda_k} , \tag{3}$$

or the process can be continued indefinitely and there are an infinite number of eigenvalues and eigenfunctions.

Note that we have denoted the least eigenvalue and the corresponding eigenfunction of $K^{(n)}(s, t)$ as λ_n and ϕ_n, which are the nth eigenvalue and the nth eigenfunction in the sequences (1) and (2). This will be justified by Theorem 2 below.

We shall examine in the next section whether the bilinear form (3) is valid for the case when the kernel $K(s, t)$ has infinite eigenvalues and eigenfunctions. The following two theorems, however, follow readily.

Theorem 1. Let the sequence $\{\phi_k(s)\}$ be all the eigenfunctions of a symmetric $\mathscr{L}_2$-kernel $K(s, t)$, with $\{\lambda_k\}$ as the corresponding eigenvalues. Then, the series

$$\sum_{n=1}^{\infty} \frac{|\phi_n(s)|^2}{\lambda_n^2}$$

converges and its sum is bounded by C_1^2, which is an upper bound of the integral

$$\int |K^2(s, t)| \, dt .$$

Proof. This result is an immediate consequence of Bessel's inequality. Indeed, the Fourier coefficients a_n of the function $K(s, t)$, with fixed s, with respect to the orthonormal system $\{\phi_n^*(s)\}$ are

$$a_n = \int K(s, t) \phi_n(t) \, dt = \phi_n(s)/\lambda_n .$$

Thus, applying Bessel's inequality, we have

$$\sum_{n=1}^{\infty} \frac{|\phi_n(s)|^2}{\lambda_n^2} \leqslant \int |K(s, t)|^2 \, dt \leqslant C_1^2 . \tag{4}$$

Theorem 2. Let the sequence $\phi_n(s)$ be all the eigenfunctions of a symmetric kernel $K(s, t)$, with $\{\lambda_n\}$ as the corresponding eigenvalues. Then, the truncated kernel

$$K^{(n+1)}(s, t) = K(s, t) - \sum_{m=1}^{n} \frac{\phi_m(s)\,\phi_m^*(t)}{\lambda_m} \tag{5}$$

has the eigenvalues $\lambda_{n+1}, \lambda_{n+2}, \ldots$, to which correspond the eigenfunctions $\phi_{n+1}(s), \phi_{n+2}(s), \ldots$ The kernel $K^{(n+1)}(s, t)$ has no other eigenvalues or eigenfunctions.

Proof. (a) Observe that the integral equation

$$\phi(s) - \lambda \int K^{(n+1)}(s, t)\,\phi(t)\,dt = 0 \tag{6}$$

is equivalent to

$$\phi(s) - \lambda \int K(s, t)\,\phi(t)\,dt + \lambda \sum_{m=1}^{n} \frac{\phi_m(s)}{\lambda_m}\,(\phi, \phi_m)\,dt = 0 \,. \tag{7}$$

If, on the left side of this equation we set $\lambda = \lambda_j$ and $\phi(s) = \phi_j(s)$ $j \geq n+1$, then, in view of the orthogonality condition, we have

$$\phi_j(s) - \lambda_j \int K(s, t)\,\phi_j(t)\,dt = 0 \,. \tag{8}$$

This means that ϕ_j and λ_j for $j \geq n+1$ are the eigenfunctions and eigenvalues of the kernel $K^{(n+1)}(s, t)$.

(b) Let λ and $\phi(s)$ be an eigenvalue and eigenfunction of the kernel $K^{(n+1)}(s, t)$ so that

$$\phi(s) - \lambda K\phi(s) + \lambda \sum_{m=1}^{n} \frac{\phi_m(s)}{\lambda_m}\,(\phi, \phi_m) = 0 \,. \tag{9}$$

Taking the scalar product of (9) with $\phi_j(s)$, $j \leq n$, we obtain

$$(\phi, \phi_j) - \lambda(K\phi, \phi_j) + (\lambda/\lambda_j)(\phi, \phi_j) = 0 \,, \tag{10}$$

where we have used the orthonormality of the ϕ_j. But $(K\phi, \phi_j) = (\phi, K\phi_j)$ $= \lambda_j^{-1}(\phi, \phi_j)$. Hence, (10) becomes

$$(\phi, \phi_j) + (\lambda/\lambda_j)\{(\phi, \phi_j) - (\phi, \phi_j)\} = (\phi, \phi_j) = 0 \,. \tag{11}$$

Thereby, the last term in the left side of equation (9) vanishes and we are left with

$$\phi(s) - \lambda \int K(s,t)\phi(t)\,dt = 0\,, \tag{12}$$

which means that λ and $\phi(s)$ are eigenvalue and eigenfunction of the kernel $K(s,t)$ and that $\phi \neq \phi_j$, $j \leqslant n$. Indeed, ϕ is orthogonal to all ϕ_j, $j \leqslant n$, and $\phi(s)$ and λ are necessarily contained in the sequences $\{\phi_k(s)\}$ and $\{\lambda_k\}$, $k \geqslant n+1$, respectively.

In light of the above two theorems, we can easily conclude that, if the symmetric kernel K has only a finite number of eigenvalues, then it is degenerate. The proof follows by observing that $K^{(n+1)}(s,t)$ then has no eigenvalues and hence it must be null. Therefore,

$$K(s,t) = \sum_{m=1}^{n} \frac{\phi_m(s)\,\phi_m{}^*(t)}{\lambda_m}\,.$$

In Chapter 2, we found that every degenerate kernel has only a finite number of eigenvalues. Combining these two results, we have the following theorem.

Theorem 3. A necessary and sufficient condition for a symmetric $\mathscr{L}_2$-kernel to be degenerate is that it have a finite number of eigenvalues.

7.4. HILBERT–SCHMIDT THEOREM AND SOME IMMEDIATE CONSEQUENCES

The pivotal result in the theory of symmetric integral equations is embodied in the following theorem.

Hilbert–Schmidt Theorem. If $f(s)$ can be written in the form

$$f(s) = \int K(s,t)h(t)\,dt\,, \tag{1}$$

where $K(s,t)$ is a symmetric $\mathscr{L}_2$-kernel and $h(t)$ is an $\mathscr{L}_2$-function, then $f(s)$ can be expanded in an absolutely and uniformly convergent Fourier series with respect to the orthonormal system of eigenfunctions (7.3.2) of the kernel K:

$$f(s) = \sum_{n=1}^{\infty} f_n \phi_n(s), \qquad f_n = (f, \phi_n).$$

The Fourier coefficients of the function $f(s)$ are related to the Fourier coefficients h_n of the function $h(s)$ by the relations

$$f_n = h_n/\lambda_n, \qquad h_n = (h, \phi_n), \tag{2}$$

where λ_n are the eigenvalues (7.3.1) of the kernel K.

Proof. The Fourier coefficients of the function $f(s)$ with respect to the orthonormal system $\{\phi_n(s)\}$ are

$$f_n = (f, \phi_n) = (Kh, \phi_n) = (h, K\phi_n) = \lambda_n^{-1}(h, \phi_n) = \lambda_n^{-1} h_n,$$

where we have used the self-adjoint property of the operator as well as the relation $\lambda_n K\phi_n = \phi_n$. Thus, the Fourier series for $f(s)$ is

$$f(s) \sim \sum_{n=1}^{\infty} f_n \phi_n(s) = \sum_{n=1}^{\infty} \frac{h_n}{\lambda_n} \phi_n(s). \tag{3}$$

The remainder term for this series can be estimated as follows:

$$\left| \sum_{k=n+1}^{n+p} h_k \frac{\phi_k(s)}{\lambda_k} \right|^2 \leq \sum_{k=n+1}^{n+p} h_k^2 \sum_{k=n+1}^{n+p} \frac{|\phi_k(s)|^2}{\lambda_k^2}$$

$$\leq \sum_{k=n+1}^{n+p} h_k^2 \sum_{k=1}^{\infty} \frac{\phi_k^2(s)}{\lambda_k^2}. \tag{4}$$

From the relation (7.3.4), we find that the above series is bounded. Also, because $h(s)$ is an $\mathscr{L}_2$-function, the series $\sum_{k=1}^{\infty} h_k^2$ is convergent and the partial sum $\sum_{k=n+1}^{n+p} h_k^2$ can be made arbitrarily small. Therefore, the series (3) converges absolutely and uniformly.

It remains to be shown that the series (3) converges to $f(s)$ in the mean. To this end, let us denote its partial sum as

$$\psi_n(s) = \sum_{m=1}^{n} \frac{h_m}{\lambda_m} \phi_m(s) \tag{5}$$

and estimate the value of $\|f(s) - \psi_n(s)\|$. Now,

$$f(s) - \psi_n(s) = Kh - \sum_{m=1}^{n} \frac{h_m}{\lambda_m} \phi_m(s) \tag{6}$$

(continued)

$$= Kh - \sum_{m=1}^{n} \frac{(h, \phi_m)}{\lambda_m} \phi_m(s) = K^{(n+1)} h , \qquad (6)$$

where $K^{(n+1)}$ is the truncated kernel as defined in the previous section. From (6), we obtain

$$\|f(s) - \psi_n(s)\|^2 = \|K^{(n+1)}h\|^2 = (K^{(n+1)}h, K^{(n+1)}h)$$
$$= (h, K^{(n+1)} K^{(n+1)}h) = (h, K_2^{(n+1)}h) , \qquad (7)$$

where we have used the self-adjointness of the kernel $K^{(n+1)}$ and also the relation $K^{(n+1)} K^{(n+1)} = K_2^{(n+1)}$. If we use property 6 of Section 7.2 and Theorem 2 of Section 7.3, we find that the least eigenvalue of the kernel $K_2^{(n+1)}$ is equal to λ_{n+1}^2. Furthermore, according to property 7 of Section 7.2, we have

$$1/\lambda_{n+1}^2 = \max[(h, K_2^{(n+1)}h)/(h,h)] , \qquad (8)$$

where we have omitted the modulus sign from the scalar product $(h, K_2^{(n+1)}h)$, as it is a positive quantity. Combining (7) and (8), we have

$$\|f(s) - \psi_n(s)\|^2 = (h, K_2^{(n+1)}h) \leqslant (h,h)/\lambda_{n+1}^2.$$

Since $\lambda_{n+1} \to \infty$, we find that $\| f(s) - \psi_n(s)\| \to 0$ as $n \to \infty$.

Finally, we use the relation

$$\|f - \psi\| \leqslant \|f - \psi_n\| + \|\psi_n - \psi\| , \qquad (9)$$

where ψ is the limit of the series with partial sum ψ_n, to prove that $f = \psi$. The first term on the right side of (9) tends to zero, as proved above. To prove that the second term also tends to zero, we observe that, since the series (3) converges uniformly, we have, for an arbitrarily small and positive ε,

$$|\psi_n(s) - \psi(s)| < \varepsilon ,$$

when n is sufficiently large. Hence, $\|\psi_n(s) - \psi(s)\| < \varepsilon(b-a)^{\frac{1}{2}}$ and the result follows.

Remark. Note that we assumed neither the convergence of the Fourier series $h(s)$ nor the completeness of the orthonormal system. We have merely used the fact that h is an $\mathscr{L}_2$-function.

An immediate consequence of the Hilbert–Schmidt theorem is the bilinear form of the type (7.3.3). Indeed, by definition,

$$M\phi = a\phi(t) - \frac{b}{\pi i}\int\limits_L^* \frac{\phi(\tau)}{\tau - t}\,d\tau \ . \tag{3}$$

From (2) and (3), it follows that

$$MLg = a\left[ag(t) + \frac{b}{\pi i}\int\limits_L^* \frac{g(t)}{\tau - t}\,d\tau \right]$$

$$- \frac{b}{\pi i}\int\limits_L^* \frac{d\tau_1}{\tau_1 - t}\left[ag(\tau_1) + \frac{b}{\pi i}\int\limits_L^* \frac{g(\tau)\,d\tau}{\tau - \tau_1} \right] = Mf \ . \tag{4}$$

Using the Poincaré–Bertrand formula (8.5.15) and after a slight simplification, equation (4) becomes

$$g(t) = \frac{a}{a^2 - b^2}f(t) - \frac{b}{(a^2 - b^2)\pi i}\int\limits_L^* \frac{f(\tau)}{\tau - t}\,d\tau \ , \tag{5}$$

where it is assumed that $a^2 - b^2 \neq 0$. Substituting (5) back in the integral equation (1), it is found that the function $g(t)$ indeed satisfies the original integral equation.

The solution of the Cauchy-type integral equation of the first kind,

$$f(t) = \frac{b}{\pi i}\int\limits_L^* \frac{g(t)}{\tau - t}\,d\tau, \tag{6}$$

follows by setting $a = 0$ in (5):

$$g(t) = \frac{1}{b\pi i}\int\limits_L^* \frac{f(\tau)}{\tau - t}\,d\tau \ . \tag{7}$$

For $b = 1$, equations (6) and (7) take the form

$$\frac{1}{\pi i}\int\limits_L^* \frac{g(\tau)\,d\tau}{\tau - t} = f(t) \ , \qquad \frac{1}{\pi i}\int\limits_L^* \frac{f(\tau)\,d\tau}{\tau - t} = g(t) \ , \tag{8}$$

which displays the reciprocity of these relations.

(ii) UNCLOSED CONTOURS AND THE RIEMANN–HILBERT PROBLEM. The analysis of the case (i) is based on the application of the Poincaré–Bertrand formula. When the contour L is not closed, then this formula is not applicable and new methods have to be devised to solve the integral equation (1). However, the Plemelj formulas (8.4.2) and (8.4.3) are valid for an arc also when we define the plus and minus directions as follows. Supplement the arc L with another arc L' so as to form a closed contour $L+L'$. Then, the interior and exterior of this closed contour stand for the plus and minus directions. Thus, we have

$$f^+(t) = \frac{1}{2}g(t) + \frac{1}{2\pi i}\int_L^* \frac{g(\tau)}{\tau-t}d\tau, \tag{9}$$

$$f^-(t) = -\frac{1}{2}g(t) + \frac{1}{2\pi i}\int_L^* \frac{g(\tau)}{\tau-t}d\tau, \tag{10}$$

for an arc L. These formulas can also be written as

$$g(t) = f^+(t) - f^-(t), \tag{11}$$

$$\frac{1}{\pi i}\int_L^* \frac{g(\tau)}{\tau-t}dt = f^+(t) + f^-(t). \tag{12}$$

Now, suppose that a function $w(t)$ is prescribed on an arc L and that it satisfies the Hölder condition on L. It is required to find a function $W(z)$ analytic for all points z on L such that it satisfies the boundary (or jump) condition

$$W^+(t) - W^-(t) = w(t), \qquad t \in L. \tag{13}$$

The formula (11) obviously helps us in evaluating such a function $W(z)$. The problem posed in (13) is a special case of the so-called Riemann–Hilbert problem, which requires the determination of a function $W(z)$ analytic for all points z not lying on L such that, for t on L,

$$W^+(t) - Z(t)W^-(t) = w(t), \tag{14}$$

where $w(t)$ and $Z(t)$ are given complex-valued functions.

It follows by substituting the formulas (11) and (12) in the integral equation

$$ag(t) = F(t) - \frac{b}{\pi i} \int\limits_L^{\!\!\!*} \frac{g(\tau)}{\tau - t} \, d\tau \,, \tag{15}$$

that the solution of this integral equation is reduced to solving the Riemann–Hilbert problem

$$(a+b)f^{+}(t) - (a-b)f^{-}(t) = F(t) \,. \tag{16}$$

We shall content ourselves with merely writing down the solution of (1). For details on the Riemann–Hilbert problem the reader is referred elsewhere [3, 13, 15].

Let L be a regular unclosed curve; then the solution of the singular integral equation (1) is

$$g(t) = \frac{a}{a^2 - b^2} f(t) - \frac{b}{(a^2 - b^2)\,\pi i} \left(\frac{t - \alpha}{t - \beta}\right)^m$$

$$\times \int\limits_L^{\!\!\!*} \left(\frac{\tau - \beta}{\tau - \alpha}\right)^m f(\tau)\frac{d\tau}{\tau - t} + \frac{c}{(t - \alpha)^{1 - m}(t - \beta)^m} \,, \tag{17}$$

where α and β are the beginning and end points of the contour L and the number m is defined as

$$m = \frac{1}{2\pi i} \log \frac{a + b}{a - b} \,.$$

The quantity c is an arbitrary constant and is suitably chosen so that $g(t)$ is bounded at α or at β.

In particular, the solution of the integral equation of the first kind (we can put $b = 1$ without any loss of generality),

$$f(t) = \frac{1}{i\pi} \int\limits_L^{\!\!\!*} \frac{g(\tau)}{\tau - t} \, d\tau \,, \tag{18}$$

is obtained from (17) by setting $a = 0$, $b = 1$. Then, $m = 1/2$ and

$$g(t) = \frac{1}{\pi i}\left(\frac{t-\alpha}{t-\beta}\right)^{\frac{1}{2}} \int_L \left(\frac{\tau-\beta}{\tau-\alpha}\right)^{\frac{1}{2}} \frac{f(\tau)}{\tau-t}\, dt + \frac{c}{[(t-\alpha)(t-\beta)]^{\frac{1}{2}}}. \quad (19)$$

8.6. THE HILBERT KERNEL

A kernel of the form

$$K(s,t) = \cot[(t-s)/2]\,,$$

where s and t are real variables, is called the Hilbert kernel and is closely connected with the Cauchy kernel. In fact, the integral equation

$$g(s) = f(s) - \lambda \int_0^{2\pi} F(s,t)\{\cot[(t-s)/2]\}g(t)\, dt\,, \quad (1)$$

where $f(s)$ and $F(s,t)$ are given continuous functions of period 2π, is equivalent to the Cauchy-type integral equation

$$g(\zeta) = f(\zeta) - \lambda \int_L [G(\zeta,\tau)/(\tau-\zeta)]g(\tau)\, d\tau\,, \quad (2)$$

where ζ and τ are complex variables and the contour L is the circumference of the unit disk with center at the point $z = 0$.

Let ζ and τ denote the points of the boundary L corresponding to the arguments s and t, respectively:

$$\zeta = e^{is}\,, \qquad \tau = e^{it}\,,$$

so that

$$\frac{d\tau}{\tau-\zeta} = i\,\frac{e^{it}\, dt}{e^{it}-e^{is}} = \left[\frac{1}{2}\cot\left(\frac{t-s}{2}\right) + \frac{i}{2}\right] dt\,. \quad (3)$$

Therefore,

$$\cot\left(\frac{t-s}{2}\right) dt = \frac{2\, d\tau}{\tau-\zeta} - \frac{d\tau}{\tau} = \frac{\tau+\zeta}{\tau-\zeta}\frac{d\tau}{\tau}\,, \quad (4)$$

and equation (1) takes the form (2).

The Hilbert kernel (1) is also related to the Poisson kernel in the integral representation formula for a harmonic function $U(r, s)$:

$$U(r, s) = \frac{1}{2\pi} \int_0^{2\pi} \frac{1-r^2}{1+r^2 - 2r\cos(t-s)} u(t)\, dt , \tag{5}$$

inside the disk $r < 1$. The function $u(t) = U(1, t)$ is the prescribed value of the harmonic function on the circumference L of the disk. Set

$$z = re^{is} , \qquad \tau = e^{it}$$

in the relation (5) and get

$$U(r, s) = \operatorname{Re}\left\{ \frac{1}{2\pi i} \int_L u(t) \frac{\tau+z}{\tau-z} \frac{d\tau}{\tau} \right\} . \tag{6}$$

Now, let $V(r, s)$ be the function that is harmonic conjugate to $U(r, s)$:

$$U(r, s) + iV(r, s) = \frac{1}{2\pi i} \int_L u(t) \frac{\tau+z}{\tau-z} \frac{d\tau}{\tau} , \tag{7}$$

such that $V(r, s)$ vanishes at the center of the disk:

$$V(r, s) \big|_{r=0} = 0 . \tag{8}$$

Then, the function $V(r, s)$ is uniquely defined.

When $r \to 1$, so that z tends to a point ζ of the circumference L from within the disk, we can apply the Plemelj formula (8.4.2) to the analytic function (7). Therefore from (7), (8.4.2), and (4), we obtain

$$v(s) = -\frac{1}{2\pi} \int_0^{*2\pi} u(t) \cot\left(\frac{t-s}{2} \right) dt , \tag{9}$$

where $v(s) = V(1, s)$ is the limiting value of the harmonic function on L. The formula (9) thus connects the limiting values of the conjugate harmonic functions $U(r, s)$ and $V(r, s)$ on the circumference.

We shall need the iterated formula formed by compounding the integrals with the Hilbert kernel:

$$g_1(s) = \frac{1}{2\pi} \int_0^{*2\pi} g(t) \cot\left(\frac{s-t}{2}\right) dt , \qquad (10)$$

and

$$g_2(s) = \frac{1}{2\pi} \int_0^{*2\pi} g_1(t) \cot\left(\frac{s-t}{2}\right) dt . \qquad (11)$$

If $U(r, s)$, $U_1(r, s)$, and $U_2(r, s)$ are the functions that are harmonic inside the disk $r < 1$, and whose values on the circumference $r = 1$ are equal respectively to $g(s)$, $g_1(s)$, and $g_2(s)$, then from (9) it follows that $U_1(r, s)$ is a harmonic function conjugate to $U(r, s)$ and $U_2(r, s)$ is conjugate to $U_1(r, s)$. The Cauchy–Riemann equations then yield the relations

$$\frac{\partial U}{\partial r} = -\frac{\partial U_2}{\partial r} , \qquad \frac{\partial U}{\partial s} = -\frac{\partial U_2}{\partial s} ,$$

or

$$U_2(r, s) = -U(r, s) + C , \qquad C = \text{const} . \qquad (12)$$

But the constant C can be determined from the condition (8):

$$C = U(r, s)\big|_{r=0} = (1/2\pi) \int_0^{2\pi} U(1, t) \, dt = (1/2\pi) \int_0^{2\pi} g(t) \, dt , \qquad (13)$$

where we have used the mean-value property of the harmonic function. Thereby, (12) becomes

$$U_2(r, s) = -U(r, s) + (1/2\pi) \int_0^{2\pi} g(t) \, dt . \qquad (14)$$

Now, put $r = 1$ in the above equation, use the relations $U(1, s) = g(s)$ and $U_2(1, s) = g_2(s)$, and get

$$g_2(s) = -g(s) + (1/2\pi) \int_0^{2\pi} g(t) \, dt . \qquad (15)$$

From (10), (11), and (15), we finally obtain the required iterated integral equation:

$$\left(\frac{1}{2\pi}\right)^2 \int\limits_0^{*2\pi} \cot\left(\frac{\sigma-s}{2}\right) d\sigma \int\limits_0^{*2\pi} g(t)\cot\left(\frac{t-\sigma}{2}\right) dt$$

$$= -g(s) + \frac{1}{2\pi}\int\limits_0^{2\pi} g(t)\, dt\,, \tag{16}$$

which is called the Hilbert formula.

8.7. SOLUTION OF THE HILBERT-TYPE SINGULAR INTEGRAL EQUATION

We can solve the integral equation of the second kind

$$ag(s) = f(s) - \frac{b}{2\pi}\int\limits_0^{*2\pi} g(t)\cot\left(\frac{t-s}{2}\right) dt\,, \tag{1}$$

where a and b are complex constants, in the same manner as we solved the corresponding Cauchy-type integral equation (8.5.1). As in Section 8.5, we define the operators L and M as

$$Lg = ag(s) + \frac{b}{2\pi}\int\limits_0^{*2\pi} g(t)\cot\left(\frac{t-s}{2}\right) dt\,, \tag{2}$$

$$M\phi = a\phi(s) - \frac{b}{2\pi}\int\limits_0^{*2\pi} \phi(t)\cot\left(\frac{t-s}{2}\right) dt\,. \tag{3}$$

Then,

$$MLg = a\left[ag(s) + \frac{b}{2\pi}\int\limits_0^{*2\pi} g(t)\cot\left(\frac{t-s}{2}\right) dt\right] \qquad \text{(equation continued)}$$

$$-\frac{b}{2\pi}\int_0^{\!*2\pi} \cot\left(\frac{t-s}{2}\right) dt$$

$$\times\left[ag(t) + \frac{b}{2\pi}\int_0^{\!*2\pi} g(\sigma)\cot\left(\frac{\sigma-t}{2}\right) d\sigma\right] = F(s) , \qquad (4)$$

where

$$F(s) = Mf = af(s) - \frac{b}{2\pi}\int_0^{\!*2\pi} f(t)\cot\left(\frac{t-s}{2}\right) dt .$$

Using the Hilbert formula (8.6.16) and simplifying, we obtain the relation

$$(a^2+b^2)g(s) - (b^2/2\pi)\int_0^{2\pi} g(t)\, dt = F(s) . \qquad (5)$$

The integral equation (5) has the simple degenerate kernel $K(s,t) = 1$, and can be readily solved by the method of Chapter 2. The result is

$$g(s) = \frac{a}{a^2+b^2}f(s) - \frac{b}{2\pi(a^2+b^2)}\int_0^{2\pi} f(t)\cot\left(\frac{t-s}{2}\right) dt$$

$$+ \frac{b^2}{2\pi a(a^2+b^2)}\int_0^{2\pi} f(t)\, dt , \qquad (6)$$

provided $a^2+b^2 \neq 0$.

For the particular case $a = 0$, the formula (6) is not applicable. Therefore, the solution of the Hilbert-type integral equation of the first kind

$$f(s) = \frac{b}{2\pi}\int_0^{\!*2\pi} g(t)\cot\left(\frac{t-s}{2}\right) dt , \qquad (7)$$

cannot be deduced from that of the second kind. But the integral equation (7) can be solved by other methods. For instance, the method of Section 8.1 is applicable here. Indeed, let us consider equation (7) with the constant b incorporated in $g(t)$:

$$f(s) = \frac{1}{2\pi} \int\limits_{0}^{*2\pi} g(t) \cot\left(\frac{t-s}{2}\right) dt . \tag{8}$$

In this equation, change s to t and t to σ, multiply both sides of the resulting equation by

$$\frac{1}{2\pi} \cot\left(\frac{t-s}{2}\right) ds ,$$

integrate from 0 to 2π, and use the Hilbert formula (8.6.16). There results the equation

$$g(s) - (1/2\pi) \int\limits_{0}^{2\pi} g(t) \, dt = F(s) , \tag{9}$$

where

$$F(s) = -\frac{1}{2\pi} \int\limits_{0}^{*2\pi} f(t) \cot\left(\frac{t-s}{2}\right) dt . \tag{10}$$

The integral equation (9) also has the simple kernel $K(s,t) = 1$ and therefore can be solved by the method of Chapter 2 if we set

$$(1/2\pi) \int\limits_{0}^{2\pi} g(s) \, ds = c .$$

Then, equation (9) becomes

$$g(s) - c = F(s) , \tag{11}$$

which, when integrated with respect to s from 0 to 2π, gives

$$\int\limits_{0}^{2\pi} F(s) \, ds = 0 . \tag{12}$$

From the relation (10), it follows that equation (12) holds for all values of the function $f(s)$. Therefore, the constant c is an arbitrary constant and the solution of the integral equation (8) is

$$g(s) = -\frac{1}{2\pi} \int\limits_{0}^{*2\pi} f(t) \cot\left(\frac{t-s}{2}\right) dt + c . \tag{13}$$

Finally, when we substitute (13) in (8), we find that the function $f(s)$ given by the relation (13) satisfies the original integral equation if, and only if,

$$\int_0^{2\pi} f(s)\, ds = 0 .\tag{14}$$

Hence, the condition (14) is necessary and sufficient for the Hilbert-type singular integral equation of the first kind to have a solution.

The second method is to use the results of Section 8.6, where we have connected the Hilbert kernel with the Cauchy kernel. For this purpose, we write $g(e^{it}) = g(t)$, etc., and assume that $g(t)$ and $f(t)$ are periodic functions with period 2π. Further, we replace $f(t)$ by $f(t)/i$. Then, the formulas (8.5.8) with the help of the transformation (8.6.3) yield the reciprocal relations

$$\frac{1}{2\pi} \!\!\int_0^{*2\pi}\!\! g(t)\cot\left(\frac{t-s}{2}\right) dt + \frac{i}{2\pi}\int_0^{2\pi} g(t)\, dt = f(s)\tag{15}$$

and

$$\frac{1}{2\pi} \!\!\int_0^{*2\pi}\!\! f(t)\cot\left(\frac{t-s}{2}\right) dt + \frac{i}{2\pi}\int_0^{2\pi} f(t)\, dt = -g(s) .\tag{16}$$

With the help of this pair of equations, the solution of the integral equation (8) can be easily deduced.

It follows from the pair (15)–(16) that, for periodic functions $f(t)$ and $g(t)$, if the condition $\int_0^{2\pi} f(t)\, dt = 0$ is satisfied, then we also have $\int_0^{2\pi} g(t)\, dt = 0$. We shall have more to say about the integral relations (15) and (16) in the next chapter.

8.8. EXAMPLES

Example 1.   Prove that the integral equation

$$F(v) = \int_0^{*1} wf(w)\frac{e^{-i\alpha(w-v)}}{w-v}\, dw \tag{1}$$

has the solution

$$wf(w) = -\frac{1}{\pi^2}\left(\frac{w}{1-w}\right)^{1/2}\int_0^{*1}\left(\frac{1-v}{v}\right)^{1/2}F(v)\frac{e^{-i\alpha(v-w)}}{v-w}\, dv. \tag{2}$$

The integral equation (1) arises in the discussion of various problems of mathematical physics. The solution follows by comparing (1) with (8.5.18) and (2) with (8.5.19), and by setting $\alpha = 0$, $\beta = 1$.

Example 2.   Solve the integral equation

$$\sum_1^\infty (a_n \cos ns + b_n \sin ns) = \frac{1}{2\pi}\int_0^{*2\pi} g(t)\cot\left(\frac{t-s}{2}\right) dt. \tag{3}$$

Observe that the function $f(s) = \sum_1^\infty (a_n \cos ns + b_n \sin ns)$ is a periodic function with period 2π. Moreover, the condition $\int_0^{2\pi} f(t)\, dt = 0$ is satisfied. Therefore, from the reciprocal pair (8.7.15) and (8.7.16), it follows that

$$g(s) = -\frac{1}{2\pi}\int_0^{*2\pi}\left[\sum_{n=1}^\infty (a_n \cos nt + b_n \sin nt)\right]\cot\left(\frac{t-s}{2}\right) dt$$

$$= \sum_{n=1}^\infty (a_n \sin ns - b_n \cos ns), \tag{4}$$

where we have left the actual integration as an exercise for the reader.

EXERCISES

Solve the following integral equations.

1. $a + bs + cs^2 + ds^3 = \int_1^s \dfrac{g(t)\, dt}{(\cos t - \cos s)^{1/2}}, \qquad 1 < s < 2,$

where a, b, c, d are real constants.

2. $\quad a + bs + cs^2 + ds^3 = \displaystyle\int_s^2 \frac{g(t)\, dt}{(\cos s - \cos t)^{1/2}}, \qquad 1 < s < 2.$

3. $\quad s^2 = \displaystyle\int_2^s \frac{g(t)\, dt}{(s^2 - t^2)^{1/3}}, \qquad 2 < s < 4.$

4. $\quad s^2 = \displaystyle\int_s^4 \frac{g(t)\, dt}{(t^2 - s^2)^{1/3}}, \qquad 2 < s < 4.$

5. $\quad as + bs^2 = \displaystyle\int_0^s \frac{g(t)\, dt}{(s - t)^{1/2}}.$

6. Substitute the solution (8.5.5) in (8.5.1) and verify that this solution satisfies the given integral equation.

7. Prove that the solution of the integral equation (8.7.5) is (8.7.6).

8. Find the solution of the integral equation

$$g(s) = (\sin s) - \frac{1}{2\pi} \int_0^{2\pi} {}^{\ast} g(t) \cot\left(\frac{t-s}{2}\right) dt.$$

9. Solve the integral equation

$$s^{-1} \int_0^s \frac{t^2 g(t)\, dt}{(s^2 - t^2)^{1/2}} = \begin{cases} -f_1(s), & 0 < s \leqslant b, \\ f_2(s), & b < s < \infty, \end{cases}$$

where b is a constant.

10. Prove that the solution of the integral equation

$$f(s) = \frac{2 s^{-2\alpha - 2\eta}}{\Gamma(\alpha)} \int_0^s t^{2\eta + 1} (s^2 - t^2)^{\alpha - 1} g(t)\, dt, \qquad 0 < \alpha < 1,$$

is

$$g(t) = \frac{t^{-2\eta-1}}{\Gamma(1-\alpha)} \frac{d}{dt} \int_0^t u^{2\alpha+2\eta+1}(t^2-u^2)^{-\alpha} f(u) \, du \; .$$

11. Prove that the solution of the integral equation

$$f(s) = \frac{2s^{2\eta}}{\Gamma(\alpha)} \int_s^\infty (t^2-s^2)^{\alpha-1} t^{-2\alpha-2\eta+1} g(t) \, dt \; , \qquad 0 < \alpha < 1 \; ,$$

is

$$g(t) = -\frac{t^{2\alpha+2\eta-1}}{\Gamma(1-\alpha)} \frac{d}{dt} \int_t^\infty u^{-2\eta+1}(u^2-t^2)^{-\alpha} f(u) \, du \; .$$

12. Solve the integral equation

$$\int_s^\infty \frac{g(t) \, dt}{(t^2-s^2)^{\frac{1}{2}}} = \begin{cases} f_1(s), & 0 \leqslant s \leqslant a, \\ -f_2(s), & a < s < \infty. \end{cases}$$

INTEGRAL TRANSFORM METHODS

9.1. INTRODUCTION

The integral transform methods are of great value in the treatment of integral equations, especially the singular integral equations. Suppose that a relationship of the form

$$g(s) = \iint \Gamma(s, x) K(x, t) g(t) \, dt \, dx \tag{1}$$

is known to be valid and that this double integral can be evaluated as an iterated integral. This means that the solution of the integral equation of the first kind,

$$f(s) = \int K(s, t) g(t) \, dt , \tag{2}$$

is

$$g(s) = \int \Gamma(s, t) f(t) \, dt . \tag{3}$$

Conversely, the relation (2) can be considered as the solution of the integral equation (3). It is conventional to refer to one of these functions as the transform of the second and to the second as an inverse transform of the first.

194

The most celebrated example of the double integral (1) is the Fourier integral

$$g(s) = (1/2\pi) \int_{-\infty}^{\infty} \int_{-\infty}^{\infty} e^{isx} e^{-ixt} g(t)\, dt\, dx \ , \tag{4}$$

which results in the reciprocal relations

$$f(s) = (2\pi)^{-\frac{1}{2}} \int_{-\infty}^{\infty} e^{-ist} g(t)\, dt \tag{5}$$

and

$$g(s) = (2\pi)^{-\frac{1}{2}} \int_{-\infty}^{\infty} e^{ist} f(t)\, dt \ . \tag{6}$$

The function $f(s)$ is known as the Fourier transform $T[g]$ of $g(t)$ and $g(s)$ as the inverse transform $T^{-1}[f]$ of $f(s)$, and vice versa. The function $f(s)$ exists if $g(t)$ is absolutely integrable, and it is square-integrable if $g(t)$ is square-integrable, as can be readily verified using Bessel's inequality. In the sequel, we shall assume that the functions involved in the integral equations as well as their transforms satisfy the appropriate regularity conditions, so that the required operations are valid.

As a second example, consider the double integral

$$g(s) = (2/\pi) \int_{0}^{\infty} \int_{0}^{\infty} (\sin sx \sin xt)\, g(t)\, dt\, dx \ . \tag{7}$$

This leads to the sine transform and its inverse,

$$f(s) = (2/\pi)^{\frac{1}{2}} \int_{0}^{\infty} (\sin st)\, g(t)\, dt \tag{8}$$

and

$$g(s) = (2/\pi)^{\frac{1}{2}} \int_{0}^{\infty} (\sin st) f(t)\, dt \ , \tag{9}$$

respectively.

For ease of notation, we shall also call the transform of f as F and that of g as G, etc., for all the transforms. It will be clear in the context as to what transforms we are implying.

9.2. FOURIER TRANSFORM

The Fourier transform $T[f]$,

$$T[f] = F(s) = (2\pi)^{-\frac{1}{2}} \int_{-\infty}^{\infty} f(t)\, e^{-ist}\, dt , \qquad (1)$$

is a linear transformation:

$$T[af+bg] = (2\pi)^{-\frac{1}{2}} \int_{-\infty}^{\infty} [af(t)+bg(t)]\, e^{-ist}\, dt$$

$$= a(2\pi)^{-\frac{1}{2}} \int_{-\infty}^{\infty} f(t)\, e^{-ist}\, dt + b(2\pi)^{-\frac{1}{2}} \int_{-\infty}^{\infty} g(t)\, e^{-ist}\, dt$$

$$= aT[f] + bT[g] . \qquad (2)$$

As such, we can use many properties of the linear operators. Furthermore, in Chapter 7 (see Exercise 11), we found that under Fourier transformation a square-integrable function preserves its norm. Hence, for such a function, we have

$$\|T[f]\| = \|F\| = \|f\| . \qquad (3)$$

Let us note some of the important properties of the Fourier transforms. They can be found in every standard book on the subject (see, for example, [17]) and, in fact, can be proved very easily by the mere use of the definitions above. These properties are:

(i) $T[f(t-a)] = e^{-ias} T[f(t)] , \qquad (4)$

where a is a constant.

(ii) $T[f(at)] = (1/|a|)\, T[f(t)]_{s \to s/a} . \qquad (5)$

(iii) $T[f'(t)] = isT[f(t)] , \qquad (6)$

where the prime denotes differentiation with respect to the argument. Similarly,

$$T[f^{k}(t)] = (is)^{k}\, T[f(t)] , \qquad (7)$$

where by $f^{k}(t)$ we mean the kth derivative of f.

(iv) If

$$h(t) = \int_a^t f(x)\, dx \,, \tag{8}$$

then

$$T[h(t)] = (1/is)\, T[f(t)] \,. \tag{9}$$

From (6) and (9), we see that the differentiation has the effect of multiplying the transform by is, whereas integration has the effect of dividing the transform by is.

(v) The convolution integral

$$h(t) = (2\pi)^{-\frac{1}{2}} \int_{-\infty}^{\infty} f(t-x)\, g(x)\, dx = (2\pi)^{-\frac{1}{2}} \int_{-\infty}^{\infty} g(t-x) f(x)\, dx \tag{10}$$

gives

$$T[h(t)] = T[f]\, T[g] \,, \tag{11}$$

or

$$H(s) = F(s)\, G(s) \,. \tag{12}$$

9.3. LAPLACE TRANSFORM

The Laplace transform $L[f]$ of a function $f(s)$ is defined as

$$L[f] = F(p) = \int_0^{\infty} f(s)\, e^{-ps}\, ds \,. \tag{1}$$

The inverse $L^{-1}[F]$ is

$$L^{-1}[F] = f(s) = (1/2\pi i) \int_{\gamma-i\infty}^{\gamma+i\infty} F(p)\, e^{ps}\, dp \,. \tag{2}$$

This transformation is also linear because

$$L[af+bg] = aL[f] + bL[g] \,,$$

for two constants a and b.

The following are some of the basic properties of the Laplace transform:

(i) $\quad F(p-a) = L[e^{as}f(s)]$, $\hspace{3cm}$ (3)

(ii) $\quad L[f(as)] = (1/a)L[f(s)]_{p\to p/a}$, $\hspace{2cm}$ (4)

(iiia) $\quad L[f'] = pL[f] - f(0)$, $\hspace{3cm}$ (5)

(b) $\quad L[f^k] = p^k L[f] - p^{k-1}f(0) - p^{k-2}f'(0) - \cdots - f^{k-1}(0)$, $\hspace{1cm}$ (6)

(c) $\quad dF(p)/dp = -L[sf(s)]$, $\hspace{3cm}$ (7)

where f^k means kth derivative with respect to the argument.

(iv) If

$$h(s) = \int_0^s f(x)\, dx ,\hspace{3cm} (8)$$

then

$$L[h] = H(p) = (1/p)L[f] .\hspace{3cm} (9)$$

(v) For the convolution integral

$$h(s) = \int_0^s f(x)\, g(s-x)\, dx = \int_0^s g(x)f(s-x)\, dx ,\hspace{1cm} (10)$$

we have

$$L[h] = L[f]L[g] ,\hspace{3cm} (11)$$

or

$$H(p) = F(p)\, G(p) ,\hspace{3cm} (12)$$

which is the same as (9.2.12).

9.4. APPLICATIONS TO VOLTERRA INTEGRAL EQUATIONS WITH CONVOLUTION-TYPE KERNELS

The basic information given here about the Fourier and Laplace transforms is sufficient to demonstrate their application to the solution of integral equations. We shall apply only the Laplace transform in this section, although the Fourier transform can also be applied just as effectively. Let us first consider the Volterra-type integral equation of the first kind,

$$f(s) = \int_0^s k(s-t)\,g(t)\,dt\,, \tag{1}$$

where $k(s-t)$ depends only on the difference $(s-t)$. Applying the Laplace transform to both sides of this equation, we obtain

$$F(p) = K(p)\,G(p)$$

or

$$G(p) = F(p)/K(p)\,. \tag{2}$$

The solution follows by inversion.

The present method is also applicable to the Volterra integral equation of the second kind with a convolution-type kernel

$$g(s) = f(s) + \int_0^s k(s-t)\,g(t)\,dt\,. \tag{3}$$

On applying Laplace transformation to both sides and using the convolution formula, we have

$$G(p) = F(p) + K(p)\,G(p)$$

or

$$G(p) = F(p)/[1-K(p)]\,, \tag{4}$$

and inversion yields the solution.

We can also find the resolvent kernel of the integral equation (3) by integral transform methods. For this purpose, we first show that, if the original kernel $k(s,t)$ is a difference kernel, then so is the resolvent kernel. Since the resolvent kernel $\Gamma(s,t)$ is a sum of the iterated kernels, all that we have to prove is that they all depend on the difference $(s-t)$. Indeed,

$$k_2(s,t) = \int_t^s k(s-x)\,k(x-t)\,dx = \int_0^{s-t} k(s-t-\sigma)\,k(\sigma)\,d\sigma\,, \tag{5}$$

where we have set $\sigma = x - t$. This process can obviously be continued and our assertion is proved. Hence, the solution of the integral equation (3) is

$$g(s) = f(s) + \int_0^s \Gamma(s-t)f(t)\,dt\ . \tag{6}$$

Application of the Laplace transform to both sides of (6) gives

$$G(p) = F(p) + \Omega(p)F(p)\ , \tag{7}$$

where

$$\Omega(p) = L[\Gamma(s-t)]\ . \tag{8}$$

From (4) and (7), we have

$$F(p)/[1-K(p)] = F(p)[1+\Omega(P)]\ , \tag{9}$$

$$\Omega(p) = K(p)/[1-K(p)]\ . \tag{10}$$

By inversion, we recover $\Gamma(s-t)$.

We illustrate the above ideas with numerous examples. Throughout these examples, we have left the evaluation of the Laplace transform and its inverse to the reader. There are numerous monographs that contain the required formulas [3, 12, 17].

9.5. EXAMPLES

Example 1. Solve the Abel integral equation

$$f(s) = \int_0^s [g(t)/(s-t)^\alpha]\,dt\ , \qquad 0 < \alpha < 1\ . \tag{1}$$

This is a convolution integral and therefore

$$F(p) = K(p)G(p)\ , \tag{2}$$

where $K(p)$ is the Laplace transform of $k(s) = s^{-\alpha}$:

$$K(p) = p^{\alpha-1}\Gamma(1-\alpha)\ . \tag{3}$$

From (2) and (3), it follows that

$$G(p) = \frac{p^{1-\alpha}F(p)}{\Gamma(1-\alpha)} = \frac{p}{\Gamma(\alpha)\Gamma(1-\alpha)} \{\Gamma(\alpha)p^{-\alpha}F(p)\}$$

$$= \frac{p}{\pi\csc\pi\alpha} \{\Gamma(\alpha)p^{-\alpha}F(p)\},\tag{4}$$

where we have used the relation $\Gamma(\alpha)\Gamma(1-\alpha) = \pi\csc\pi\alpha$. Now if we use the relation (9.3.12), (4) becomes

$$G(p) = \frac{\sin\alpha\pi}{\pi} pL\left[\int_0^s (s-t)^{\alpha-1}f(t)\,dt\right].\tag{5}$$

By virtue of the property (9.3.5), we finally have

$$g(s) = \frac{\sin\alpha\pi}{\pi}\frac{d}{ds}\int_0^s (s-t)^{\alpha-1}f(t)\,dt,\tag{6}$$

which agrees with the relation (8.1.4) obtained in the previous chapter by a different method.

Example 2. Solve the integral equation

$$s = \int_0^s e^{s-t}g(t)\,dt.\tag{7}$$

Taking the Laplace transform of both sides, we obtain

$$1/p^2 = K(p)G(p),\tag{8}$$

where $K(p)$ is the Laplace transform of $k(s) = e^s$:

$$K(p) = \int_0^\infty e^s e^{-sp}\,ds = 1/(p-1).\tag{9}$$

The result of combining (7), (8), and (9) is

$$G(p) = (p-1)/p^2 = (1/p) - (1/p^2),$$

whose inverse is

$$g(s) = 1 - s . \tag{10}$$

Example 3.   Solve the integral equation

$$\sin s = \int_0^s J_0(s-t) g(t) \, dt . \tag{11}$$

Here, the function $k(s) = J_0(s)$, whose Laplace transform is known to be $1/(1+p^2)^{1/2}$. Also, the Laplace transform of $\sin s$ is $1/(1+p^2)$. Therefore, when we take the Laplace transform of equation (11), there results the relation

$$G(p) = 1/(1+p^2)^{1/2} ,$$

which by inversion yields the solution

$$g(s) = J_0(s) . \tag{12}$$

Incidently, by substituting (12) back in (11), we get the interesting result

$$\int_0^s J_0(s-t) J_0(t) \, dt = \sin s . \tag{13}$$

Example 4.   Recall that we solved integral equations of the type

$$f(s) = \int_0^s k(s^2 - t^2) g(t) \, dt , \qquad s > 0 , \tag{14}$$

in the previous chapter for some special cases of the kernel $k(s^2 - t^2)$. With the help of the Laplace transform, we can solve equation (14) for a general convolution kernel. For this purpose, the first step is to set

$$s = u^{1/2} , \qquad t = \sigma^{1/2} , \qquad g_1(\sigma) = \tfrac{1}{2}\sigma^{-1/2} g(\sigma^{1/2}) , \qquad f_1(u) = f(u^{1/2}) . \tag{15}$$

Then, the integral equation (14) takes the form

$$f_1(u) = \int_0^u k(u-\sigma) g_1(\sigma) \, d\sigma , \qquad u > 0 . \tag{16}$$

Taking the Laplace transform of both sides of (16), we get

$$G_1(p) = F_1(p)/K(p) = pF_1(p)/pK(p) . \tag{17}$$

By defining

$$1/pK(p) = H(p) , \tag{18}$$

the relation (17) becomes

$$G_1(p) = pH(p) F_1(p) . \tag{19}$$

Using the relations (9.3.5) and (9.3.12), we have

$$G_1(p) = L\left[\frac{d}{du} \int_0^u h(u-\sigma)f_1(\sigma) \, d\sigma\right] \tag{20}$$

or

$$g_1(u) = \frac{d}{du} \int_0^u h(u-\sigma)f_1(\sigma) \, d\sigma , \tag{21}$$

where $h(s)$ stands for the inverse of the function $H(p)$. Finally, from (15) and (21), we have the required solution

$$g(s) = 2\frac{d}{ds} \int_0^s tf(t) h(s^2 - t^2) \, dt . \tag{22}$$

Let us solve (14) for some special cases:

(a) $k(t) = t^{-\alpha}$, $0 < \alpha < 1$. This means that we have to solve the integral equation

$$f(s) = \int_0^s [g(t)/(s^2 - t^2)^\alpha] \, dt . \tag{23}$$

The solution follows from (22) if we can evaluate the function $h(s)$ from the relation

$$H(p) = 1/pK(p) . \tag{24}$$

But

$$K(p) = \int_0^\infty t^{-\alpha} e^{-pt} \, dt = \Gamma(1-\alpha) p^{\alpha-1} \, .$$

Therefore,

$$h(s) = L^{-1} \left[\frac{1}{p^\alpha \Gamma(1-\alpha)} \right] = \frac{\sin \alpha \pi}{\pi} s^{\alpha-1} \, ,$$

and

$$g(s) = \frac{2 \sin \alpha \pi}{\pi} \frac{d}{ds} \int_0^s \frac{tf(t) \, dt}{(s^2 - t^2)^{1-\alpha}} \, , \tag{25}$$

which agrees with the relation (8.2.9).

(b) $k(t) = t^{-\frac{1}{2}} \cos(\beta t^{\frac{1}{2}})$, where β is a constant. In this case, $K(p) = \pi^{\frac{1}{2}} p^{-\frac{1}{2}} \exp(-\beta^2/4p)$. Therefore,

$$h(s) = L^{-1} [\pi^{-\frac{1}{2}} p^{-\frac{1}{2}} \exp(\beta^2/4p)] = \pi^{-1} s^{-\frac{1}{2}} \cosh(\beta t^{\frac{1}{2}}) \, . \tag{26}$$

This means that the solution of the integral equation

$$f(s) = \int_0^s \frac{\cos[\beta(s^2 - t^2)^{\frac{1}{2}}]}{(s^2 - t^2)^{\frac{1}{2}}} g(t) \, dt \, , \qquad s > 0 \, , \tag{27}$$

is

$$g(s) = \frac{2}{\pi} \frac{d}{ds} \int_0^s \frac{\cosh[\beta(s^2 - t^2)^{\frac{1}{2}}]}{(s^2 - t^2)^{\frac{1}{2}}} tf(t) \, dt \, . \tag{28}$$

Note that the relations (27) and (28) remain valid for $0 < s < \infty$.

Example 5.   Solve the inhomogeneous integral equation

$$g(s) = 1 - \int_0^s (s-t) g(t) \, dt \, . \tag{29}$$

Since

$$k(s) = s \, , \qquad K(p) = 1/p^2 \, , \tag{30}$$

the application of the Laplace transform gives

$$G(p) = (1/p) - [G(p)/p^2]$$

or

$$G(p) = p/(1+p^2) = L[\cos s] .$$

Hence, the solution is

$$g(s) = \cos s . \tag{31}$$

Example 6.   Find the resolvent of the integral equation

$$g(s) = f(s) + \int_0^s (s-t)g(t)\, dt . \tag{32}$$

Here again $k(s) = s$, and we have $K(p) = 1/p^2$. The formula (9.4.10) gives $\Omega(p) = 1/(p^2 - 1)$, whose inverse is $\Gamma(s) = \frac{1}{2}(e^s - e^{-s})$. Therefore, the value of the resolvent kernel is

$$\Gamma(s-t) = \tfrac{1}{2}(e^{s-t} - e^{-s+t}) ,$$

and the solution of the integral equation (32) is

$$g(s) = f(s) + \tfrac{1}{2}e^s \int_0^s e^{-t}f(t)\, dt - \tfrac{1}{2}e^{-s} \int_0^s e^t f(t)\, dt . \tag{33}$$

Example 7.   Find the resolvent of the integral equation

$$g(s) = f(s) + \int_0^s e^{s-t} g(t)\, dt . \tag{34}$$

Here, $k(s) = e^s$, which gives $K(p) = 1/(p-1)$, and from the formula (9.4.10), we have $\Omega(p) = 1/(p-2)$, $\Gamma(s) = e^{2s}$. Hence, the resolvent kernel is $\Gamma(s-t) = e^{2s-2t}$, and the solution of the integral equation (34) is

$$g(s) = f(s) + \int_0^s e^{2s-2t} f(t)\, dt . \tag{35}$$

Example 8.   Solve the integral equation

$$g(s) = f(s) + \lambda \int_0^s J_0(s-t)g(t)\, dt . \tag{36}$$

The kernel $k(s) = \lambda J_0(s)$, and therefore

$$K(p) = \frac{\lambda}{(1+p^2)^{\frac{1}{2}}}, \qquad \Omega(p) = \frac{\lambda}{(1+p^2)^{\frac{1}{2}} - \lambda},$$

$$\Gamma(s) = \frac{\lambda}{(1-\lambda^2)^{\frac{1}{2}}} \int_0^s [\sin(1-\lambda^2)^{\frac{1}{2}}](s-\sigma)\frac{J_1(\sigma)}{\sigma}\, d\sigma$$

$$+ \lambda\{\cos[(1-\lambda^2)^{\frac{1}{2}}s]\} + \frac{\lambda^2}{(1-\lambda^2)^{\frac{1}{2}}}\sin[(1-\lambda^2)^{\frac{1}{2}}s]. \tag{37}$$

The value of the resolvent kernel follows by setting $(s-t)$ for s and the solution of the integral equation (35) is then readily obtained from the formula (9.4.6).

Example 9. As a final example, solve the inhomogeneous Abel integral equation:

$$g(s) = f(s) + \lambda \int_0^s [g(t)/(s-t)^\alpha]\, dt, \qquad 0 < \alpha < 1. \tag{38}$$

The kernel $k(s) = \lambda s^{-\alpha}$ yields

$$K(p) = \lambda\Gamma(1-\alpha)p^{\alpha-1},$$
$$\Omega(p) = \lambda\Gamma(1-\alpha)p^{\alpha-1}/[1 - \lambda\Gamma(1-\alpha)p^{\alpha-1}]. \tag{39}$$

The inverse of (39) is

$$\Gamma(s) = \sum_{n=1}^\infty \frac{[\lambda\Gamma(1-\alpha)s^{1-\alpha}]^n}{s\Gamma[n(1-\alpha)]}. \tag{40}$$

Hence, the solution of the integral equation (38) is

$$g(s) = f(s) + \int_0^s \frac{\sum\limits_{n=1}^\infty [\lambda\Gamma(1-\alpha)(s-t)^{1-\alpha}]^n}{(s-t)\Gamma[n(1-\alpha)]} f(t)\, dt. \tag{41}$$

9.6. HILBERT TRANSFORM

The finite Hilbert transform of a function $g(\varphi)$ is usually defined as

$$f(\theta) = \frac{1}{\pi} \int_0^{*\pi} \frac{\sin \theta}{\cos \theta - \cos \varphi} g(\varphi) \, d\varphi \, , \qquad (1)$$

with the inverse

$$g(\theta) = \frac{1}{\pi} \int_0^{*\pi} \frac{\sin \varphi}{\cos \varphi - \cos \theta} f(\varphi) \, d\varphi + \frac{1}{\pi} \int_0^{\pi} g(\varphi) \, d\varphi \, . \qquad (2)$$

Various other forms of the Hilbert transform pair can be deduced from equations (1) and (2). In this connection, we need the relation

$$\int_0^{*\pi} \frac{\cos n\varphi \, d\varphi}{\cos \varphi - \cos \alpha} = \pi \frac{\sin n\alpha}{\sin \alpha} \, , \qquad (3)$$

which can be proved by induction since the cases $n = 0$ and $n = 1$, are elementary relations.

From (1) and (2), it is apparent that $f(-\theta) = -f(\theta)$, $g(-\theta) = g(\theta)$. Now, set

$$f_1(\theta) = -f(\theta) \, , \qquad \text{so that} \quad f_1(-\theta) = -f_1(\theta) \, , \qquad (4)$$

$$(1/\pi) \int_0^\pi g(\varphi) \, d\varphi = C \, , \qquad C = \text{const} \, , \qquad (5)$$

$$g_1(\theta) = g(\theta) - C \, , \qquad \text{so that} \quad g_1(-\theta) = g(\theta) - C \, . \qquad (6)$$

From the above relations, it follows that

$$(1/\pi) \int_0^\pi g_1(\theta) \, d\theta = (1/\pi) \int_0^\pi g(\theta) \, d\theta - (1/\pi) \int_0^\pi C \, d\theta = 0 \, , \qquad (7)$$

and

$$\frac{1}{\pi} \int_0^{*\pi} \frac{\sin \theta}{\cos \varphi - \cos \theta} g_1(\varphi) \, d\varphi = \frac{1}{\pi} \int_0^{*\pi} \frac{\sin \theta}{\cos \varphi - \cos \theta} [g(\varphi) - C] \, d\varphi$$

$$= -f(\theta) - \frac{C}{\pi} \int\limits_{0}^{*\pi} \frac{\sin\theta}{\cos\varphi - \cos\theta} \, d\varphi = f_1(\theta). \tag{8}$$

Writing $\theta = \frac{1}{2}(\theta - \varphi) + \frac{1}{2}(\theta + \varphi)$ in equation (8), we get

$$f_1(\theta) = (1/2\pi) \int\limits_{0}^{*\pi} [\cot\tfrac{1}{2}(\theta + \varphi)] g_1(\varphi) \, d\varphi$$

$$+ (1/2\pi) \int\limits_{0}^{*\pi} [\cot\tfrac{1}{2}(\theta - \varphi)] g_1(\varphi) \, d\varphi. \tag{9}$$

In the first integral, replace φ by $-\varphi$ and use the relation (6). Combine the resulting integral with the second integral in (9). The result is

$$f_1(\theta) = (1/2\pi) \int\limits_{-\pi}^{*\pi} [\cot\tfrac{1}{2}(\theta - \varphi)] g_1(\varphi) \, d\varphi. \tag{10}$$

Similarly, by starting with equation (2) and going through the same steps as above, we get the relation

$$g_1(\theta) = (1/2\pi) \int\limits_{0}^{*\pi} [\cot\tfrac{1}{2}(\varphi - \theta)] f_1(\varphi) \, d\varphi$$

$$+ (1/2\pi) \int\limits_{0}^{*\pi} [\cot\tfrac{1}{2}(\varphi + \theta)] f_1(\varphi) \, d\varphi. \tag{11}$$

After replacing φ by $-\varphi$ in the second integral and using the relation (4), we obtain from (11)

$$g_1(\theta) = (1/2\pi) \int\limits_{-\pi}^{*\pi} [\cot\tfrac{1}{2}(\varphi - \theta)] f_1(\varphi) \, d\varphi. \tag{12}$$

The relations (10) and (12) constitute a second form of the finite Hilbert transform pair.

The third form of the Hilbert transform pair can be deduced from (9) and (11). In fact, the relation (9) can be written as

$$f_1(\theta) = (1/2\pi) \int\limits_{0}^{*\pi} [1 + \cot\tfrac{1}{2}(\theta + \varphi)] g_1(\varphi) \, d\varphi$$

$$+ (1/2\pi) \int_0^{*\pi} [1 + \cot\tfrac{1}{2}(\theta - \varphi)] g_1(\varphi)\, d\varphi - (1/\pi) \int_0^{\pi} g_1(\varphi)\, d\varphi .$$

$$(13)$$

The last integral vanishes because of (7), while the first integral can be combined with the second by replacing φ by $-\varphi$ and using (6). The result is

$$f_1(\theta) = (1/2\pi) \int_{-\pi}^{*\pi} [1 + \cot\tfrac{1}{2}(\theta - \varphi)] g_1(\varphi)\, d\varphi .$$

$$(14)$$

Similarly, the relation (11) takes the form

$$g_1(\theta) = (1/2\pi) \int_{-\pi}^{*\pi} [1 + \cot\tfrac{1}{2}(\varphi - \theta)] f_1(\varphi)\, d\varphi .$$

$$(15)$$

The transform pair (14)–(15) is precisely the reciprocal pair of Hilbert-type singular integral equations encountered in Section 8.7 of the previous chapter except for a trivial adjustment of the symbols and the range of integration.

A fourth form of the finite Hilbert transform pair, which is nonangular, is obtained from the pair (1)–(2) by setting

$$x = \cos\theta , \qquad y = \cos\varphi , \qquad p(x) = \frac{f(\theta)}{\sin\theta} = \frac{f(\cos^{-1}x)}{(1-x^2)^{\frac{1}{2}}} ,$$

$$q(x) = \frac{g(\theta)}{\sin\theta} = \frac{g(\cos^{-1}x)}{(1-x^2)^{\frac{1}{2}}} .$$

$$(16)$$

Then, equation (1) becomes

$$\frac{f(\theta)}{\sin\theta} = \frac{1}{\pi} \int_0^{*\pi} \frac{1}{\cos\theta - \cos\varphi} \frac{g(\varphi)}{\sin\varphi} \sin\varphi\, d\varphi$$

or

$$p(x) = \frac{1}{\pi} \int_{-1}^{*1} \frac{q(y)\, dy}{x - y} , \qquad -1 < x < 1 ,$$

$$(17)$$

the so-called airfoil equation. Similarly, (2) takes the form

$$q(x) = \frac{1}{\pi} \int_{-1}^{*1} \left(\frac{1-y^2}{1-x^2}\right)^{\frac{1}{2}} \frac{p(y)}{y-x} \, dy + \frac{C}{(1-x^2)^{\frac{1}{2}}}, \qquad -1 < x < 1, \quad (18)$$

where

$$C = (1/\pi) \int_{-1}^{1} q(y) \, dy$$

has the character of an arbitrary constant.

The pair of equations (17)–(18) is a special case of the pair of integral equations (8.5.18)–(8.5.19).

The infinite Hilbert transform is defined as

$$f(s) = (1/\pi) \int_{-\infty}^{*\infty} [g(t)/(t-s)] \, dt . \qquad (19)$$

Its inverse is

$$g(s) = -(1/\pi) \int_{-\infty}^{*\infty} [f(t)/(t-s)] \, dt . \qquad (20)$$

9.7. EXAMPLES

Example 1. Solve the homogeneous integral equation

$$\int_{-1}^{*1} [g(y)/(x-y)] \, dy = 0 . \qquad (1)$$

The solution follows from (9.6.18):

$$g(x) = C/(1-x^2)^{\frac{1}{2}} . \qquad (2)$$

Example 2. Solve the integral equation

$$\sin s = (1/\pi) \int_{-\infty}^{*\infty} [g(t)/(t-s)] \, dt . \qquad (3)$$

To solve this equation, let us consider the integral

$$\int_{-\infty}^{*\infty} [e^{it}/(s-t)]\, dt = \pi i \sum (\text{residues of the poles on the } t \text{ axis})$$

$$= \pi i (\cos s + i \sin s) . \tag{4}$$

Separating the real and imaginary parts, we obtain

$$(1/\pi) \int_{-\infty}^{*\infty} [(\cos t)/(s-t)]\, dt = -\sin s , \tag{5}$$

$$(1/\pi) \int_{-\infty}^{*\infty} [(\sin t)/(s-t)]\, dt = \cos s . \tag{6}$$

Comparing (3) and (5), we have the solution

$$g(t) = \cos t . \tag{7}$$

EXERCISES

1. Show that the solution of the integral equation

$$f(s) = 2 \int_s^1 \frac{t g(t)\, dt}{(t^2 - s^2)^{1/2}}$$

is

$$g(s) = -\frac{1}{\pi s} \frac{d}{ds} \int_s^1 \frac{t f(t)\, dt}{(t^2 - s^2)^{1/2}} .$$

Find the solution for the following two special cases: (i) $f(s) = 2s^2/(1-s^2)^{1/2}$; (ii) $f(s) = s^2$.

2. Solve the integral equation

$$f(s) = s \int_s^\infty \frac{g'(t)}{(t^2 - s^2)^{1/2}}\, dt .$$

3. Solve the Abel integral equation of the second kind

$$g(s) = s^{-\frac{1}{2}} e^{-a/4s} + \frac{i}{\sqrt{\pi}} \int_0^s \frac{g(t)}{(s-t)^{\frac{1}{2}}} \, dt \, .$$

This integral equation arises in the theory of wave propagation over a flat surface.

4. If it is required that in the Hilbert transform pair (9.6.17)–(9.6.18) the function $q(-1)$ be finite, show that there must follow

$$\int_{-1}^{1} q(t) \, dt = \int_{-1}^{\bullet 1} (1-t^2)^{\frac{1}{2}} p(t) \frac{dt}{1+t} \, ,$$

and verify that in this case the solution $q(s)$ becomes

$$q(s) = \frac{1}{\pi} \left(\frac{1+s}{1-s} \right)^{\frac{1}{2}} \int_{-1}^{\bullet 1} \left(\frac{1-t}{1+t} \right)^{\frac{1}{2}} \frac{p(t)}{s-t} \, dt \, .$$

5. With the help of finite Hilbert transform, solve the equation

$$s^2 = \int_{-\ell}^{\ell} \frac{2t g(t)}{s^2 - t^2} \, dt \, , \qquad \text{assuming that} \qquad g(t) = -g(-t).$$

6. With the help of finite Hilbert transform, solve the equation

$$as + b + \sigma_\ell (\log |\ell - s|) - \sigma_0 \log |s| = \int_0^\ell [g'(t)/(t-s)] \, dt$$

subject to the conditions

$$g'(0) = \sigma_0 \, , \qquad g'(\ell) = \sigma_\ell \, , \qquad g(0) = g_0 \, , \qquad g(\ell) = g_\ell \, .$$

7. Use the formula

$$\int_0^\infty t^{\alpha-1} \cos st \, dt = \Gamma(\alpha) |s|^{-\alpha} \cos(\tfrac{1}{2}\alpha\pi) \, , \qquad 0 < \alpha < 1 \, ,$$

and show that the integral equation

$$f(s) = \int_0^1 [g(t)/|s-t|^\alpha]\, dt$$

has no more than one solution.

8. Solve the integral equation

$$\sum_{r=0}^n a_r s^r = \frac{1}{\pi} \int_0^{*\ell} \frac{g(t)\, dt}{t-s},$$

where a_r are given constants.

 Hint: Make the substitutions

$$s = (\ell/2)(1 - \cos\theta), \qquad t = (\ell/2)(1 - \cos\varphi).$$

9. Use the method of Section 9.4 and find the resolvent for the integral equation

$$g(s) = f(s) + \int_0^s (s^2 - t^2) g(t)\, dt.$$

10. Solve the integral equation

$$g(s) = f(s) + \lambda \int_0^s J_1(s-t) g(t)\, dt.$$

11. Find the resolvent of the integral equation

$$g(s) = f(s) + \int_0^s \frac{(s-t)^{m-1}}{(m-1)!} g(t)\, dt,$$

and complete the solution of the Example 3 in Section 5.3.

12. Use the infinite Hilbert transform pair and solve the integral equation

$$1/(1+s^2) = \int_{-\infty}^{*\infty} [g(t)/(s-t)]\, dt.$$

APPLICATIONS TO MIXED BOUNDARY VALUE PROBLEMS

Mixed boundary value problems occur in physical sciences rather frequently and various mathematical techniques have been used to solve them. In this chapter, we present an integral-equation method applicable to most of these problems.

10.1. TWO-PART BOUNDARY VALUE PROBLEMS

An integral equation of the form

$$\int_0^a K_0(t, \rho) g(t) \, dt = f(\rho) , \qquad 0 < \rho < a , \qquad (1)$$

where the function $f(\rho)$ and the kernel $K_0(t, \rho)$ are known and $g(t)$ is to be evaluated, embodies the solution of various mixed boundary value problems in potential theory, elastostatics, steady heat conduction, the flow of perfect fluids, and various other problems of equilibrium states. The boundaries involved are those of solids such as circular disks, elliptic disks, spherical caps, and spheroidal caps.

The integral equation (1) is a Fredholm integral equation of the first kind and is therefore, in general, difficult to solve. However, it is possible to reduce the solution of (1) to that of a pair of Volterra integral equations of the first kind with rather simple kernels. This reduction is achieved for every kernel $K_0(t, \rho)$ that for all $g(t)$ satisfies the relation

$$\int_0^a K_0(t, \rho) g(t) \, dt = h_1(\rho) \int_0^\rho K_2(w, \rho) [h_2(w)]^2$$

$$\times \int_w^a K_2(w, t) g(t) h_3(t) \, dt \, dw, \qquad 0 < \rho < a, \quad (2)$$

where h_1, h_2, h_3, and K_2 are known functions. It is further assumed that the kernel K_2 is such that the Volterra integral equations

$$\int_0^\rho K_2(t, \rho) g(t) \, dt = f(\rho), \qquad 0 < \rho < a, \tag{3}$$

and

$$\int_\rho^a K_2(\rho, t) g(t) \, dt = f(\rho), \qquad 0 < \rho < a \tag{4}$$

possess explicit unique solutions for g in terms of f, for all arbitrary differentiable functions f.

The task of solving (1) is now readily accomplished if we define two functions $S(\rho)$ and $C(\rho)$ such that

$$S(\rho) = h_2(\rho) \int_\rho^a K_2(\rho, t) g(t) h_3(t) \, dt, \qquad 0 < \rho < a, \tag{5}$$

and

$$f(\rho) = h_1(\rho) \int_0^\rho K_2(w, \rho) C(w) h_2(w) \, dw, \qquad 0 < \rho < a. \tag{6}$$

With the help of relations (2), (5), and (6), equation (1) takes the form

$$h_1(\rho) \int_0^\rho K_2(w, \rho) h_2(w) S(w) \, dw = h_1(\rho) \int_0^\rho K_2(w, \rho) C(w) h_2(w) \, dw, \quad (7)$$

$$0 < \rho < a,$$

or

$$S(\rho) = C(\rho), \qquad 0 < \rho < a. \tag{8}$$

In view of the assumptions already made about the solutions of the Volterra integral equations (3) and (4), we can solve equation (6) for the function $C(\rho)$ and hence, from (8), $S(\rho)$ is known. We can then invert the integral equation (5) and obtain the required function $g(t)$. We illustrate the above analysis with the following example.

Example. Solve the integral equation

$$\int_0^a t\phi(t) \int_0^\infty J_1(p\rho)J_1(pt)\, dp\, dt = \Omega\rho, \qquad 0 < \rho < a, \tag{9}$$

where $\phi(t)$ is the unknown function. This equation solves the problem of the torsion of an isotropic and homogeneous elastic half-space due to a uniformly rotating, rigid circular disk which is attached to its free face (see Section 6.6, Example 2). The function $J_1(x)$ is the Bessel function. Comparing (1) and (9), we have

$$g(t) = t\phi(t), \qquad f(\rho) = \Omega\rho, \qquad K_0(t,\rho) = \int_0^\infty J_1(p\rho)J_1(pt)\, dp. \tag{10}$$

The kernel K_0 satisfies the relation (2) because, for all $g(t)$, we can write

$$\int_0^a K_0(t,\rho)g(t)\, dt = \int_0^a g(t) \int_0^\infty J_1(p\rho)J_1(pt)\, dp\, dt$$

$$= \int_0^a g(t) \int_0^\infty \frac{2p}{\pi\rho t} \int_0^\rho \int_0^t \frac{J_{\frac{1}{2}}(pw)J_{\frac{1}{2}}(pv)(wv)^{\frac{3}{2}}\, dv\, dw\, dp\, dt}{(\rho^2 - w^2)^{\frac{1}{2}}(t^2 - v^2)^{\frac{1}{2}}}$$

$$= \frac{2}{\pi\rho} \int_0^a t^{-1} g(t) \int_0^\rho \int_0^t \frac{\delta(w-v)(wv)\, dv\, dw\, dt}{(\rho^2 - w^2)^{\frac{1}{2}}(t^2 - v^2)^{\frac{1}{2}}}$$

$$= \frac{2}{\pi\rho} \int_0^a t^{-1} g(t) \int_0^{\min(\rho,t)} \frac{w^2 \, dw \, dt}{(\rho^2-w^2)^{\frac{1}{2}}(t^2-w^2)^{\frac{1}{2}}}$$

$$= \frac{2}{\pi\rho} \int_0^\rho \frac{w^2}{(\rho^2-w^2)^{\frac{1}{2}}} \int_w^a \frac{t^{-1} g(t) \, dt \, dw}{(t^2-w^2)^{\frac{1}{2}}}, \qquad 0 < \rho < a, \qquad (11)$$

where we have used the first Sonine integral:

$$J_n(p\rho) = \left(\frac{2p}{\pi}\right)^{\frac{1}{2}} \frac{1}{\rho^n} \int_0^\rho \frac{J_{n-(\frac{1}{2})}(pw) \, w^{n+(\frac{1}{2})}}{(\rho^2-w^2)^{\frac{1}{2}}} \, dw, \qquad (12)$$

and the relation

$$\int_0^\infty p J_\mu(pw) J_\mu(pv) \, dp = \delta(w-v)/(wv)^{\frac{1}{2}}, \qquad (13)$$

with δ the Dirac delta function. We have further used the sifting property of this function and changed the order of integration as explained in Figure 10.1.

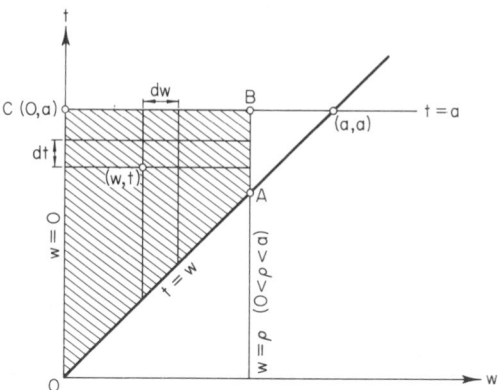

Figure 10.1

Comparing (2) and (11), we obtain the values of the functions h_1, h_2, h_3, and K_2 as

$$h_1(\rho) = 2/\pi\rho, \qquad h_2(\rho) = \rho, \qquad h_3(\rho) = 1/\rho,$$

$$K_2(t,\rho) = (\rho^2 - t^2)^{-\frac{1}{2}}. \tag{14}$$

Furthermore, the kernel K_2 is simple enough to ensure the inversion of the integral equations (3) and (4) (see Example 3 of Section 8.2). The present method is therefore applicable. Indeed, we set

$$S(\rho) = \rho \int_\rho^a \frac{\phi(t)\,dt}{(t^2 - \rho^2)^{\frac{1}{2}}}, \tag{15}$$

$$\Omega\rho = \frac{2}{\pi\rho} \int_0^\rho \frac{wS(w)\,dw}{(\rho^2 - w^2)^{\frac{1}{2}}}, \tag{16}$$

and the integral equation (9) is identically satisfied.

Finally, we invert (16) and obtain

$$S(\rho) = \frac{\Omega}{\rho} \frac{d}{d\rho} \int_0^\rho \frac{t^3\,dt}{(\rho^2 - t^2)^{\frac{1}{2}}} = 2\Omega\rho, \tag{17}$$

and then (15) yields the value of the function $\phi(\rho)$ as

$$\phi(\rho) = -\frac{4\Omega}{\pi} \frac{d}{d\rho} \int_\rho^a \frac{u\,du}{(u^2 - \rho^2)^{\frac{1}{2}}} = \frac{4\Omega\rho}{\pi(a^2 - \rho^2)^{\frac{1}{2}}}. \tag{18}$$

10.2. THREE-PART BOUNDARY VALUE PROBLEMS

A three-part boundary value problem has an integral representation formula of the form

$$\int_b^a K_0(t,\rho)\,g(t)\,dt = f(\rho)\,, \qquad b < \rho < a\,, \tag{1}$$

where b and a are two given numbers such as the inner and outer radii of an annular disk or the bounding angles of an annular spherical cap. The function f and the kernel K_0 are known, while g is to be determined. Let us set

$$f(\rho) = \sum_{r=-\infty}^{\infty} a_r \rho^r = f_1(\rho) + f_2(\rho)\,, \tag{2}$$

where

$$f_1(\rho) = \sum_{r=0}^{\infty} a_r \rho^r\,, \qquad 0 \leqslant \rho < a\,, \tag{3}$$

and

$$f_2(\rho) = \sum_{r=-\infty}^{-1} a_r \rho^r\,, \qquad b < \rho < \infty\,. \tag{4}$$

In addition, we define two functions $g_1(\rho)$ and $g_2(\rho)$ such that

$$g_1(\rho) + g_2(\rho) = \begin{cases} 0\,, & 0 \leqslant \rho < b\,, \\ g(\rho)\,, & b \leqslant \rho \leqslant a\,, \\ 0\,, & a < \rho < \infty\,. \end{cases} \tag{5}$$

From the relations (2)–(5), it follows that the integral equation (1) splits into two integral equations:

$$\int_0^\infty K_0(t,\rho)\,g_1(t)\,dt = f_1(\rho)\,, \qquad 0 < \rho < a\,, \tag{6}$$

and

$$\int_0^\infty K_0(t,\rho)\,g_2(t)\,dt = f_2(\rho)\,, \qquad b < \rho < \infty\,. \tag{7}$$

Proceeding as in Section 2, we assume that the kernel $K_0(t,\rho)$ is such that for all $g(t)$ it satisfies the relation

$$\int_0^\infty K_0(t,\rho)\,g(t)\,dt = \begin{cases} h_{11}(\rho)\displaystyle\int_0^\rho K_2(w,\rho)\,[h_{12}(w)]^2 \\[2ex] \quad\times \displaystyle\int_w^\infty K_2(w,t)\,g(t)\,h_{13}(t)\,dt\,dw, \qquad 0 < \rho < \infty, \\[3ex] h_{21}(\rho)\displaystyle\int_\rho^\infty K_2(\rho,w)\,[h_{22}(w)]^2 \\[2ex] \quad\times \displaystyle\int_0^w K_2(t,w)\,g(t)\,h_{23}(t)\,dt\,dw, \qquad 0 < \rho < \infty, \end{cases} \tag{8}$$

where h_{ij} ($i = 1, 2$; $j = 1, 2, 3$), and the kernel K_2 are known functions. Moreover, the kernel K_2 is such that the Volterra integral equations

$$\int_0^\rho K_2(t,\rho)\,g(t)\,dt = f(\rho), \qquad 0 < \rho < \infty, \tag{9}$$

and

$$\int_\rho^\infty K_2(\rho,t)\,g(t)\,dt = f(\rho), \qquad 0 < \rho < \infty, \tag{10}$$

possess unique solutions for g in terms of all arbitrary differentiable functions f.

From relation (6) and the first part of relation (8), we have

$$h_{11}(\rho)\int_0^\rho K_2(w,\rho)\,[h_{12}(w)]^2 \int_w^\infty K_2(w,t)\,g_1(t)\,h_{13}(t)\,dt\,dw = f_1(\rho), \tag{11}$$

$$0 < \rho < a.$$

Similarly, relation (7) and the second part of (8) give

$$h_{21}(\rho)\int_\rho^\infty K_2(\rho,w)\,[h_{22}(w)]^2 \int_0^w K_2(t,w)\,g_2(t)\,h_{23}(t)\,dt\,dw = f_2(\rho), \tag{12}$$

$$b < \rho < \infty.$$

The next step is to define unknown functions $S_1, S_2, T_1, T_2, C_1,$ and C_2 such that

$$h_{12}(\rho) \int_{\rho}^{\infty} K_2(\rho, t) g_1(t) h_{13}(t) \, dt = \begin{cases} S_1(\rho), & 0 < \rho < a, \\ -T_1(\rho), & a < \rho < \infty, \end{cases} \tag{13}$$

$$h_{22}(\rho) \int_{0}^{\rho} K_2(t, \rho) g_2(t) h_{23}(t) \, dt = \begin{cases} -T_2(\rho), & 0 < \rho < b, \\ S_2(\rho), & b < \rho < \infty, \end{cases} \tag{14}$$

$$h_{11}(\rho) \int_{0}^{\rho} K_2(w, \rho) C_1(w) h_{12}(w) \, dw = f_1(\rho), \qquad 0 < \rho < a, \tag{15}$$

$$h_{21}(\rho) \int_{\rho}^{\infty} K_2(\rho, w) C_2(w) h_{22}(w) \, dw = f_2(\rho), \qquad b < \rho < \infty. \tag{16}$$

These Volterra-type integral equations are similar to equations (9) and (10), whose solutions are assumed to be known. From (11), (13), and (15), we derive the equation

$$h_{11}(\rho) \int_{0}^{\rho} K_2(w, \rho) h_{12}(w) S_1(w) \, dw$$

$$= h_{11}(\rho) \int_{0}^{\rho} K_2(w, \rho) C_1(w) h_{12}(w) \, dw, \qquad 0 < \rho < a, \tag{17}$$

or

$$S_1(\rho) = C_1(\rho), \qquad 0 < \rho < a. \tag{18}$$

Similarly, the result of combining (12), (14), and (16) is

$$S_2(\rho) = C_2(\rho), \qquad b < \rho < \infty. \tag{19}$$

The functions C_1 and C_2 can be evaluated in terms of the known functions f_1 and f_2 from equations (15) and (16). Hence, S_1 and S_2 are known. That leaves two unknown functions T_1 and T_2 still to be evaluated. For this purpose, we appeal to relations (5) and (13). The result is

$$h_{12}(\rho) \int_{\rho}^{\infty} K_2(\rho, t) g_2(t) h_{13}(t) \, dt = T_1(\rho), \qquad a < \rho < \infty. \tag{20}$$

Similarly, from (5) and (14), we have

$$h_{22}(\rho) \int_0^\rho K_2(t,\rho) g_1(t) h_{23}(t)\, dt = T_2(\rho), \qquad 0 < \rho < b. \tag{21}$$

Now, invert (14) to find the value of $g_2(t)$ in terms of T_2 and S_2 and substitute this value of $g_2(t)$ in (20). There results an integral equation containing the unknown functions T_1 and T_2. Likewise, the relations (13) and (21) lead to a second integral equation for T_1 and T_2. Both these equations are Fredholm integral equations of the second kind, and can therefore be solved by a straightforward iterative method.

Example.   Solve the integral equation

$$\int_b^a t\phi(t) \int_0^\infty J_1(p\rho) J_1(pt)\, dp\, dt = \Omega\rho, \qquad b < \rho < a, \tag{22}$$

which embodies the solution of the torsion of an isotropic and homogeneous elastic half-space due to a uniformly rotating annular disk with inner radius b and outer radius a. Comparing it with (1) and (5), we have

$$g(t = t\phi(t), \qquad g_1(t) = t\phi_1(t), \qquad g_2(t) = t\phi_2(t), \tag{23}$$

$$f(\rho) = \Omega\rho, \qquad f_1(\rho) = \Omega\rho, \qquad 0 \leqslant \rho < a; \qquad f_2(\rho) = 0, \tag{24}$$
$$b < \rho < \infty;$$

$$K_0(t,\rho) = \int_0^\infty J_1(p\rho) J_1(pt)\, dp, \tag{25}$$

where

$$\phi_1(\rho) + \phi_2(\rho) = \begin{cases} 0, & 0 \leqslant \rho < b, \\ \phi(\rho), & b \leqslant \rho \leqslant a, \\ 0, & a < \rho < \infty. \end{cases} \tag{26}$$

In addition, the kernel $K_0(t,\rho)$ satisfies the requirement (8) in as much as, for all $g(t)$, we have

$$\int_0^\infty K_0(t,\rho) g(t)\, dt = \int_0^\infty g(t) \int_0^\infty J_1(p\rho) J_1(pt)\, dp\, dt$$

$$= \int_0^\infty g(t) \int_0^\infty \frac{2p}{\pi \rho t} \int_0^\rho \int_0^t \frac{J_{\frac{1}{2}}(pw) J_{\frac{1}{2}}(pv) (wv)^{3/2} \, dv \, dw \, dp \, dt}{(\rho^2 - w^2)^{1/2} (t^2 - v^2)^{1/2}}$$

$$= \frac{2}{\pi \rho} \int_0^\infty t^{-1} g(t) \int_0^\rho \int_0^t \frac{\delta(w-v) (wv) \, dv \, dw \, dt}{(\rho^2 - w^2)^{1/2} (t^2 - v^2)^{1/2}}$$

$$= \frac{2}{\pi \rho} \int_0^\infty t^{-1} g(t) \int_0^{\min(\rho, t)} \frac{w^2 \, dw \, dt}{(\rho^2 - w^2)^{1/2} (t^2 - w^2)^{1/2}}$$

$$= \frac{2}{\pi \rho} \int_0^\rho \frac{w^2}{(\rho^2 - w^2)^{1/2}} \int_w^\infty \frac{t^{-1} g(t) \, dt \, dw}{(t^2 - w^2)^{1/2}}, \qquad 0 < \rho < \infty, \quad (27)$$

and

$$\int_0^\infty K_0(t, \rho) g(t) \, dt = \int_0^\infty g(t) \int_0^\infty J_1(p\rho) J_1(pt) \, dp \, dt$$

$$= \int_0^\infty g(t) \int_0^\infty \frac{2p\rho t}{\pi} \int_\rho^\infty \int_t^\infty \frac{J_{\frac{3}{2}}(pw) J_{\frac{3}{2}}(pv) \, dv \, dw \, dp \, dt}{(wv)^{1/2} (w^2 - \rho^2)^{1/2} (v^2 - t^2)^{1/2}}$$

$$= \frac{2\rho}{\pi} \int_0^\infty tg(t) \int_\rho^\infty \int_t^\infty \frac{\delta(w-v) \, dv \, dw \, dt}{(wv) (w^2 - \rho^2)^{1/2} (v^2 - t^2)^{1/2}}$$

$$= \frac{2\rho}{\pi} \int_0^\infty tg(t) \int_{\max(\rho, t)}^\infty \frac{w^{-2} \, dw \, dt}{(w^2 - \rho^2)^{1/2} (w^2 - t^2)^{1/2}}$$

$$= \frac{2\rho}{\pi} \int_\rho^\infty \frac{w^{-2}}{(w^2 - \rho^2)^{1/2}} \int_0^w \frac{tg(t) \, dt \, dw}{(w^2 - t^2)^{1/2}}, \qquad 0 < \rho < \infty, \quad (28)$$

where we have used the formulas (10.1.12) and (10.1.13) and the equation

$$J_n(p\rho) = \left(\frac{2p}{\pi}\right)^{1/2} \rho^n \int_\rho^\infty \frac{J_{n+(1/2)}(pw) w^{-[n-(1/2)]}}{(w^2 - \rho^2)^{1/2}} \, dw \,.$$

Figure 10.2

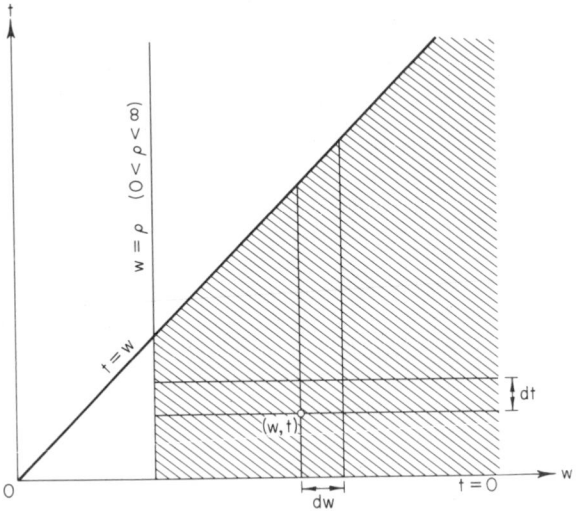

Figure 10.3

Furthermore, we have changed the order of integration in the steps leading to formulas (27) and (28) as explained in Figures 10.2 and 10.3. Hence,

$$h_{11}(\rho) = 2/\pi\rho\,, \qquad h_{12}(\rho) = \rho\,, \qquad h_{13}(\rho) = 1/\rho\,,$$

$$h_{21}(\rho) = 2\rho/\pi\,, \qquad h_{22}(\rho) = 1/\rho\,, \qquad h_{23}(\rho) = \rho\,, \qquad (29)$$

$$K_2(t,\rho) = (\rho^2 - t^2)^{-\frac{1}{2}}\,.$$

Furthermore, the kernel K_2 is such that the Volterra integral equations (9) and (10) can be readily solved, and therefore the method of this section can be applied.

The system of integral equations that corresponds to the system (13)–(21) is

$$\rho \int_{\rho}^{\infty} \frac{\phi_1(t)\,dt}{(t^2 - \rho^2)^{\frac{1}{2}}} = \begin{cases} S_1(\rho)\,, & 0 < \rho < a\,, \\ -T_1(\rho)\,, & a < \rho < \infty\,, \end{cases} \qquad (30)$$

$$\rho^{-1} \int_{0}^{\rho} \frac{t^2\,\phi_2(t)\,dt}{(\rho^2 - t^2)^{\frac{1}{2}}} = \begin{cases} -T_2(\rho)\,, & 0 < \rho < b\,, \\ S_2(\rho)\,, & b < \rho < \infty\,, \end{cases} \qquad (31)$$

$$\frac{2}{\pi\rho} \int_{0}^{\rho} \frac{wS_1(w)\,dw}{(\rho^2 - w^2)^{\frac{1}{2}}} = \Omega\rho\,, \qquad 0 < \rho < a\,, \qquad (32)$$

$$\frac{2\rho}{\pi} \int_{\rho}^{\infty} \frac{S_2(w)\,dw}{w(w^2 - \rho^2)^{\frac{1}{2}}} = 0\,, \qquad b < \rho < \infty\,, \qquad (33)$$

$$\rho \int_{\rho}^{\infty} \frac{\phi_2(t)\,dt}{(t^2 - \rho^2)^{\frac{1}{2}}} = T_1(\rho)\,, \qquad a < \rho < \infty\,, \qquad (34)$$

$$\rho^{-1} \int_{0}^{\rho} \frac{t^2\,\phi_1(t)\,dt}{(\rho^2 - t^2)^{\frac{1}{2}}} = T_2(\rho)\,, \qquad 0 < \rho < b\,. \qquad (35)$$

The integral equations (30)–(33) are readily inverted and the results are (see Example 3 of Section 8.2 and Example 4 of Section 9.5)

$$\phi_1(\rho) = -\frac{2}{\pi}\frac{d}{d\rho}\left[\int_{\rho}^{a} \frac{S_1(u)\,du}{(u^2 - \rho^2)^{\frac{1}{2}}} - \int_{a}^{\infty} \frac{T_1(u)\,du}{(u^2 - \rho^2)^{\frac{1}{2}}}\right]\,, \qquad (36)$$

$$\phi_2(\rho) = \frac{2}{\pi\rho^2}\frac{d}{d\rho}\left[-\int_0^b \frac{u^2\,T_2(u)\,du}{(\rho^2-u^2)^{1/2}} + \int_b^\rho \frac{u^2\,S_2(u)\,du}{(\rho^2-u^2)^{1/2}}\right], \qquad (37)$$

$$S_1(\rho) = \frac{\Omega}{\rho}\frac{d}{d\rho}\int_0^\rho \frac{t^3\,dt}{(\rho^2-t^2)^{1/2}} = 2\Omega\rho, \qquad (38)$$

$$S_2(\rho) = 0. \qquad (39)$$

Substituting these values in (34) and (35), we get

$$T_1(\rho) = \frac{2\rho}{\pi}\int_0^b u^2\,T_2(u)\int_\rho^\infty \frac{t^{-1}\,dt\,du}{(t^2-\rho^2)^{1/2}(t^2-u^2)^{3/2}}$$

$$= \frac{1}{\rho(\pi)^{1/2}\,\Gamma(5/2)}\int_0^b \frac{u^2\,T_2(u)\,{}_2F_1(1/2,1;5/2;u^2/\rho^2)\,du}{(\rho^2-u^2)}, \qquad (40)$$

$$a < \rho < \infty,$$

and

$$T_2(\rho) = \frac{4\Omega}{\pi\rho}\int_0^\rho \frac{t^3\,dt}{(\rho^2-t^2)^{1/2}(a^2-t^2)^{1/2}} + \frac{2}{\pi\rho}\int_0^\rho \frac{t^3}{(\rho^2-t^2)^{1/2}}\int_a^\infty \frac{T_1(u)\,du\,dt}{(u^2-t^2)^{3/2}}$$

$$= \frac{8\Omega\rho^2\,a^3}{3\pi(\rho^2+a^2)^2}\,{}_2F_1\left(1,\frac{3}{2};\frac{5}{2};\frac{4a^2\,\rho^2}{(\rho^2+a^2)^2}\right)$$

$$+ \frac{\rho^2}{(\pi)^{1/2}\,\Gamma(5/2)}\int_a^\infty \frac{T_1(u)\,{}_2F_1(1/2,1;5/2;\rho^2/u^2)\,du}{u(u^2-\rho^2)}, \qquad (41)$$

$$0 < \rho < b.$$

In the above relations, ${}_2F_1$ stands for the hypergeometric function and we have used the following relations pertaining to this function:

$$\int_\rho^\infty \frac{t^{-1}\,dt}{(t^2-\rho^2)^{1/2}(t^2-u^2)^{3/2}} = \frac{\pi^{1/2}\rho^{-2}}{2\Gamma(5/2)\,(\rho^2-u^2)}\,{}_2F_1\left(\frac{1}{2},1;\frac{5}{2};\frac{u^2}{\rho^2}\right),$$

$$u < \rho,$$

$$\int_0^\rho \frac{t^3\,dt}{(\rho^2-t^2)^{\frac12}(u^2-t^2)^{\frac32}} = \frac{\pi^{\frac12}\rho^3\,{}_2F_1(1/2,1;5/2;\rho^2/u^2)}{2\Gamma(5/2)\,u(u^2-\rho^2)}\,,\qquad \rho < u\,,$$

$$\int_0^\rho \frac{t^3\,dt}{(\rho^2-t^2)^{\frac12}(a^2-t^2)^{\frac12}} = \frac12\frac{\pi^{\frac12}(a\rho)^3}{\Gamma(5/2)(\rho^2+a^2)^2}\,{}_2F_1\!\left(1,\frac32;\frac52;\frac{4a^2\rho^2}{(\rho^2+a^2)^2}\right),$$

$$\rho < a\,.$$

Equations (40) and (41) are two simultaneous Fredholm integral equations of the second kind and can be solved approximately by iteration when we introduce the parameter $\lambda = b/a$, such that $\lambda \ll 1$. Indeed, the hypergeometric function ${}_2F_1$ occurring under the integral signs in these equations is reducible to an elementary function:

$$_2F_1\!\left(\frac12,1;\frac52;\frac{x^2}{y^2}\right) = \frac{3y}{4x^3}\left[2xy-(y^2-x^2)\log\!\left(\frac{y+x}{y-x}\right)\right],\qquad x < y.\quad(42)$$

Thereby equations (40) and (41) take the simple forms

$$T_1(a\rho) = \frac1\pi\int_0^1 T_2(bu)\left[\frac{2\lambda\rho}{\rho^2-\lambda^2u^2} - \frac1u\log\!\left(\frac{\rho+\lambda u}{\rho-\lambda u}\right)\right]du\,,\qquad(43)$$

$$1 < \rho < \infty\,,$$

and

$$T_2(b\rho) = \frac{8\Omega\lambda^2\rho^2 a}{3\pi(1+\lambda^2\rho^2)^2}\,{}_2F_1\!\left(1,\frac32;\frac52;\frac{4\lambda^2\rho^2}{(1+\lambda^2\rho^2)^2}\right)$$

$$+\frac1{\lambda\rho\pi}\int_1^\infty T_1(au)\left[\frac{2\lambda u\rho}{u^2-\lambda^2\rho^2} - \log\!\left(\frac{u+\lambda\rho}{u-\lambda\rho}\right)\right]du\,,$$

$$0 < \rho < 1\,.\qquad(44)$$

We first attend to (44) and observe that

$$_2F_1\!\left(1,\frac32;\frac52;x^2\right) = 1 + 3\sum_{n=1}^\infty \frac{x^{2n}}{2n+3}\,.\qquad(45)$$

With the help of this value, we readily obtain the first iteration for T_2 as

$$T_2(b\rho) = \frac{8\Omega a}{3\pi} \left\{ \lambda^2 \rho^2 + \frac{2\rho^4 \lambda^4}{5} + O(\lambda^6) \right\}. \tag{46}$$

This value in turn helps us in solving (43) approximately

$$T_1(a\rho) = \frac{32\Omega a\lambda^5}{45\pi^2} \left[\frac{1}{\rho^3} \left(1 + \frac{2}{7}\lambda^2 \right) + \frac{6\lambda^2}{7\rho^5} + O(\lambda^4) \right]. \tag{47}$$

In the above approximations, we have included only those terms that are needed to evaluate the torque experienced by the annulus up to $O(\lambda^9)$.

Finally, we substitute the values of S_1, S_2, T_1, and T_2 as given by the relations (38), (39), (46), and (47) in (36) and (37), and get

$$\phi_1(\rho) = \frac{4\Omega}{\pi} \left(\frac{\rho/a}{[1 - (\rho^2/a^2)]^{1/2}} + \frac{16\lambda^5}{45\pi^2} \left\{ \left(1 + \frac{2\lambda^2}{7} \right) \frac{a^3}{2\rho^3} \right. \right.$$

$$\times \left[-\frac{3a}{\rho} \sin^{-1} \frac{\rho}{a} + \left(1 - \frac{\rho^2}{a^2} \right)^{1/2} + 2 \left(1 - \frac{\rho^2}{a^2} \right)^{-1/2} \right] + \frac{3\lambda^2}{28} \frac{a^5}{\rho^5}$$

$$\times \left[-15 \frac{a}{\rho} \sin^{-1} \frac{\rho}{a} - 2 \left(1 - \frac{\rho^2}{a^2} \right)^{3/2} + 9 \left(1 - \frac{\rho^2}{a^2} \right)^{1/2} \right.$$

$$\left. \left. \left. + 8 \left(1 - \frac{\rho^2}{a^2} \right)^{-1/2} \right] \right\} + O(\lambda^9) \right), \tag{48}$$

$$\phi_2(\rho) = -\frac{4\Omega}{3\pi^2} \left\{ 2\lambda \left[\frac{3\rho}{b} \sin^{-1} \frac{b}{\rho} - \left(1 - \frac{b^2}{\rho^2} \right)^{1/2} - 2 \left(1 - \frac{b^2}{\rho^2} \right)^{-1/2} \right] \right.$$

$$+ \frac{\lambda^3}{5} \frac{\rho^2}{b^2} \left[\left(\frac{15\rho}{b} \right) \sin^{-1} \frac{b}{\rho} + 2 \left(1 - \frac{b^2}{\rho^2} \right)^{3/2} - 9 \left(1 - \frac{b^2}{\rho^2} \right)^{1/2} \right.$$

$$\left. \left. - 8 \left(1 - \frac{b^2}{\rho^2} \right)^{-1/2} \right] + O(\lambda^5) \right\}. \tag{49}$$

Substituting these values in the relation $\phi(\rho) = \phi_1(\rho) + \phi_2(\rho)$, we obtain the desired solution of the integral equation (1).

10.3. GENERALIZED TWO-PART BOUNDARY VALUE PROBLEMS

An integral equation of a more general type such as

$$\int_0^a g(t) K_1(t, \rho)\, dt = f(\rho), \qquad 0 < \rho < a, \tag{1}$$

where the kernels K_1 can be perturbed on the kernel $K_0(t, \rho)$ of Section 10.1, can also be solved by the present method. This necessitates the splitting of the kernel K_1 as

$$K_1(t, \rho) = K_0(t, \rho) + G(t, \rho), \tag{2}$$

where the kernel $G(t, \rho)$ is in some sense smaller than K_0. From (1) and (2), it follows that

$$\int_0^a K_0(t, \rho) g(t)\, dt = f(\rho) - \int_0^a G(t, \rho) g(t)\, dt, \qquad 0 < \rho < a. \tag{3}$$

The kernel $K_0(t, \rho)$ satisfies the same requirements as those in Section 10.1. From (10.1.2) and (3), we have

$$h_1(\rho) \int_0^\rho K_2(w, \rho) [h_2(w)]^2 \int_w^a g(t) h_3(t) K_2(w, t)\, dt\, dw$$

$$= f(\rho) - \int_0^a G(t, \rho) g(t)\, dt, \qquad 0 < \rho < a. \tag{4}$$

Now we attempt to put the right side of this equation in the form of the left side as we did in equation (10.1.7). This is done by defining two functions $S(\rho)$ and $C(\rho)$ as in (10.1.5) and (10.1.6) and a new function $L(v, w)$ such that

$$G(t, \rho) = h_1(\rho) h_3(t) \int_0^\rho \int_0^t K_2(w, \rho) K_2(v, t) h_2(w) h_2(v) L(v, w)\, dv\, dw. \tag{5}$$

Thus, the integral on the right side of (4) takes the form

$$\int_0^a G(t, \rho) g(t)\, dt = \int_0^a g(t) h_1(\rho) h_3(t) \qquad \text{(equation continued)}$$

$$\times \int_0^\rho \int_0^t K_2(w,\rho)\, K_2(v,t)\, h_2(w)\, h_2(v)\, L(v,w)\, dv\, dw\, dt$$

$$= h_1(\rho) \int_0^\rho K_2(w,\rho)\, h_2(w) \int_0^a L(v,w)\, h_2(v)$$

$$\times \int_v^a K_2(v,t)\, g(t)\, h_3(t)\, dt\, dv\, dw\,, \qquad 0 < \rho < a\,, \quad (6)$$

where we have assumed that various orders of integration can be interchanged. When we substitute (10.1.5), (10.1.6), and (6) in equation (4), we get

$$h_1(\rho) \int_0^\rho K_2(w,\rho)\, h_2(w)\, S(w)\, dw = h_1(\rho) \int_0^\rho K_2(w,\rho)\, C(w)\, h_2(w)\, dw$$

$$- h_1(\rho) \int_0^\rho K_2(w,\rho)\, h_2(w)$$

$$\times \int_0^a L(v,w)\, S(v)\, dv\, dw\,, \qquad (7)$$

$$0 < \rho < a\,.$$

From this equation, it follows that

$$S(\rho) = C(\rho) - \int_0^a L(v,\rho)\, S(v)\, dv\,, \qquad 0 < \rho < a\,, \qquad (8)$$

which is a Fredholm integral equation of the second kind and can be solved for $S(\rho)$. The required function $g(t)$ is then obtained by inverting (10.1.5).

Example. Solve the integral equation

$$\int_0^a t\phi(t) \int_0^\infty [pJ_1(p\rho)J_1(pt)/\gamma]\, dp\, dt = \Omega\rho\,, \qquad 0 < \rho < a\,, \qquad (9)$$

where

$$\gamma = \begin{cases} -i(k^2 - p^2)^{\frac{1}{2}}\,, & k \geqslant p\,, \\ (p^2 - k^2)^{\frac{1}{2}}\,, & p \geqslant k\,. \end{cases} \qquad (10)$$

As explained in Section 6.6, Example 2, this equation solves the problem of the torsional oscillations of an isotropic and homogeneous elastic half-space due to a rigid circular disk of radius a which is performing simple harmonic oscillations.

Comparison of (9) and (1) gives

$$g(t) = t\phi(t), \qquad f(\rho) = \Omega\rho, \qquad (11)$$

$$K_1(t, \rho) = \int_0^\infty [pJ_1(p\rho)J_1(pt)/\gamma]\, dp .$$

We split K_1 as in (2) with

$$K_0(t, \rho) = \int_0^\infty J_1(p\rho)J_1(pt)\, dp , \qquad (12)$$

$$G(t, \rho) = \int_0^\infty \left(\frac{p}{\gamma} - 1\right) J_1(p\rho)J_1(pt)\, dp$$

$$= \frac{2}{\pi\rho t} \int_0^\rho \int_0^t \frac{(wv)^{3/2}}{(\rho^2 - w^2)^{1/2}(t^2 - w^2)^{1/2}}$$

$$\times \left[\int_0^\infty p\left(\frac{p}{\gamma} - 1\right) J_{1/2}(pv)J_{1/2}(pw)\, dp\right] dv\, dw . \qquad (13)$$

By following the method of this section and borrowing the known results in Section 10.1, we have

$$h_1(\rho) = 2/\pi\rho , \qquad h_2(\rho) = \rho , \qquad h_3(\rho) = 1/\rho ,$$

$$K_2(t, \rho) = (\rho^2 - t^2)^{-1/2} , \qquad (14)$$

$$L(v, w) = (vw)^{1/2} \int_0^\infty [p(p/\gamma - 1)] J_{1/2}(pv)J_{1/2}(pw)\, dp , \qquad (15)$$

$$S(\rho) = \rho \int_\rho^a \frac{\phi(t)\, dt}{(t^2 - \rho^2)^{1/2}} , \qquad (16)$$

$$\Omega\rho = \frac{2}{\pi\rho} \int_0^\rho \frac{wC(w)\,dw}{(\rho^2 - w^2)^{\frac{1}{2}}}, \tag{17}$$

and

$$S(\rho) = C(\rho) - \int_0^a L(v,\rho)\,S(v)\,dv, \qquad 0 < \rho < a. \tag{18}$$

Equation (17) is readily inverted as shown in relation (10.1.17):

$$C(\rho) = 2\Omega\rho. \tag{19}$$

Therefore, equation (18) becomes

$$S(\rho) = 2\Omega\rho - \int_0^a L(v,\rho)\,S(v)\,dv, \qquad 0 < \rho < a. \tag{20}$$

The infinite integral (15) can be converted to a finite integral (see Appendix A.2):

$$L(v,w) = \begin{cases} i(vw)^{\frac{1}{2}} \int_0^k [p^2/(k^2 - p^2)^{\frac{1}{2}}]\, H_{\frac{1}{2}}{}^{(1)}(pv)\, J_{\frac{1}{2}}(pw)\,dp, \\[2pt] \qquad v \geqslant w, \\[12pt] i(vw)^{\frac{1}{2}} \int_0^k [p^2/(k^2 - p^2)^{\frac{1}{2}}]\, J_{\frac{1}{2}}(pv)\, H_{\frac{1}{2}}{}^{(1)}(pw)\,dp, \\[2pt] \qquad w \geqslant v, \end{cases} \tag{21}$$

where $H_{\frac{1}{2}}{}^{(1)}$ is a Hankel function of the first kind. This form of the kernel is useful for small values of k.

With this much information, we can solve the integral equation (20) approximately for small values of ak, which happens to be a dimensionless parameter. For this purpose, we write it as

$$S(a\rho) = 2\Omega a\rho - a \int_0^1 L(av, a\rho)\,S(av)\,dv, \qquad 0 < \rho < 1. \tag{22}$$

An approximate value of the kernel $aL(av, a\rho)$ is obtained from (21) by using the series expansions for the Hankel and Bessel functions. The result is

$$aL(av, a\rho) = \begin{cases} \dfrac{\alpha^2 \rho}{2} + \dfrac{4i\alpha^3 \rho v}{3\pi} - \dfrac{\alpha^4}{16}(3\rho v^2 + \rho^3) - \dfrac{8i\alpha^5}{45\pi}(\rho^3 v + \rho v^3) \\[2mm] \quad + \dfrac{\alpha^6}{384}(5\rho v^4 + 10\rho^3 v^2 + \rho^5) \\[2mm] \quad + \dfrac{4i\alpha^7}{1575\pi}(3\rho v^5 + 10\rho^3 v^3 + 3\rho^5 v) + O(\alpha^8), \\[2mm] \qquad v \geqslant \rho, \\[5mm] \dfrac{\alpha^2 v}{2} + \dfrac{4i\alpha^3 \rho v}{3\pi} - \dfrac{\alpha^4}{16}(3\rho^2 v + v^3) - \dfrac{8i\alpha^5}{45\pi}(\rho v^3 + \rho^3 v) \\[2mm] \quad + \dfrac{\alpha^6}{384}(5\rho^4 v + 10\rho^2 v^3 + v^5) \\[2mm] \quad + \dfrac{4i\alpha^7}{1575\pi}(3\rho v^5 + 10\rho^3 v^3 + 3\rho^5 v) + O(\alpha^8), \\[2mm] \qquad \rho \geqslant v, \end{cases} \tag{23}$$

where $x = ak$. By applying the straightforward iteration method to (22), we obtain an approximate value for $S(a\rho)$ as

$$S(a\rho) = 2\Omega a [c_1(\alpha)\rho + c_3(\alpha)\rho^3 + c_5(\alpha)\rho^5 + c_7(\alpha)\rho^7 + O(\alpha^8)], \tag{24}$$

where

$$c_1(\alpha) = 1 - \frac{\alpha^2}{4} - \frac{4i\alpha^3}{9\pi} + \frac{19\alpha^4}{192} + \frac{53i\alpha^5}{225\pi} - \left(\frac{16}{81\pi^2} + \frac{143}{3840}\right)\alpha^6 - \frac{8051i\alpha^7}{58800\pi},$$

$$c_3(\alpha) = \frac{\alpha^2}{12} + \frac{\alpha^4}{96} + \frac{i\alpha^5}{45\pi} - \frac{11\alpha^6}{2304} - \frac{47i\alpha^7}{4200\pi},$$

$$c_5(\alpha) = -\frac{\alpha^4}{960} - \frac{\alpha^6}{3840} - \frac{i\alpha^7}{1680\pi}, \qquad c_7(\alpha) = \frac{\alpha^6}{80640}.$$

Finally, we invert (16) to get

$$\phi(\rho) = -\frac{2}{\pi}\frac{d}{d\rho}\int_\rho^a \frac{S(u)\,du}{(u^2 - \rho^2)^{1/2}} = -\frac{2}{\pi}\frac{d}{d\rho}\int_{\rho/a}^1 \frac{S(au)\,du}{[u^2 - (\rho^2/a^2)]^{1/2}}. \tag{25}$$

From (24) and (25), we have

$$\phi(\rho) = \frac{4\Omega\rho/a}{\pi[1-(\rho^2/a^2)]^{1/2}} \left\{ 1 - \alpha^2 \left[\frac{1}{6} + \frac{1}{6}\left(1 - \frac{\rho^2}{a^2}\right) \right] - \frac{4i\alpha^3}{9\pi} \right.$$

$$+ \alpha^4 \left[\frac{13}{120} - \frac{1}{60}\left(1 - \frac{\rho^2}{a^2}\right) - \frac{1}{360}\left(1 - \frac{\rho^2}{a^2}\right)^2 \right]$$

$$+ \frac{i\alpha^5}{\pi} \left[\frac{58}{225} - \frac{2}{45}\left(1 - \frac{\rho^2}{a^2}\right) \right] - \alpha^6 \left[\left(\frac{16}{81\pi^2} + \frac{17}{1680} \right) \right.$$

$$\left. - \frac{53}{5040}\left(1 - \frac{\rho^2}{a^2}\right) + \frac{1}{1680}\left(1 - \frac{\rho^2}{a^2}\right)^2 + \frac{1}{25200}\left(1 - \frac{\rho^2}{a^2}\right)^3 \right]$$

$$\left. - \frac{i\alpha^7}{\pi} \left[\frac{1093}{7350} - \frac{13}{525}\left(1 - \frac{\rho^2}{a^2}\right) + \frac{1}{630}\left(1 - \frac{\rho^2}{a^2}\right)^2 \right] + O(\alpha^8) \right\}. \quad (26)$$

When $\alpha \to 0$, this reduces to equation (10.1.18).

10.4. GENERALIZED THREE-PART BOUNDARY VALUE PROBLEMS

Finally, we consider the integral equation

$$\int_b^a K_1(t,\rho) g(t) \, dt = f(\rho), \qquad b < \rho < a, \qquad (1)$$

which is the generalization of the integral equation (10.2.1) and the kernel K_1 is to be perturbed on K_0 of Section 10.2. Indeed, we split it as

$$K_1(t,\rho) = K_0(t,\rho) + G(t,\rho), \qquad (2)$$

and assume that $G(t,\rho)$ is in some sense smaller than K_0. With the help of the relations (10.2.2)–(10.2.5) and (2), the integral equation (1) becomes equivalent to the pair of equations

$$\int_0^\infty K_0(t,\rho) g_1(t) \, dt = f_1(\rho) - \int_0^\infty G(t,\rho) g_1(t) \, dt, \qquad 0 < \rho < a, \quad (3)$$

$$\int_0^\infty K_0(t,\rho)\,g_2(t)\,dt = f_2(\rho) - \int_0^\infty G(t,\rho)\,g_2(t)\,dt\,, \qquad b < \rho < \infty\,. \quad (4)$$

Furthermore, the choice of the kernel K_0 is such that the requirements embodied in the relations (10.2.8)–(10.2.10) are satisfied.

Now we extend the analysis of Section 10.3 and define two new kernels $L_1(v, w)$ and $L_2(v, w)$ such that

$$G(t,\rho) = \begin{cases} h_{11}(\rho)h_{13}(t) \displaystyle\int_0^\rho \int_0^t K_2(w,\rho)\,K_2(v,t)\,h_{12}(w)\,h_{12}(v) \\[2pt] \qquad \times L_1(v, w)\,dv\,dw\,, \\[10pt] h_{21}(\rho)h_{23}(t) \displaystyle\int_\rho^\infty \int_t^\infty K_2(\rho,w)\,K_2(t,v)\,h_{22}(w)\,h_{22}(v) \\[2pt] \qquad \times L_2(v, w)\,dv\,dw\,, \end{cases} \quad (5)$$

where the h's and the kernel K_2 are the same functions as occur in (10.2.8). Thus, the integrals on the right side of the relations (3) and (4) take the forms

$$\int_0^\infty G(t,\rho)\,g_1(t)\,dt = \int_0^\infty g_1(t)\,h_{11}(\rho)\,h_{13}(t)$$

$$\times \int_0^\rho \int_0^t K_2(w,\rho)\,K_2(v,t)\,h_{12}(w)\,h_{12}(v)$$

$$\times L_1(v, w)\,dv\,dw\,dt\,,$$

$$= h_{11}(\rho) \int_0^\rho K_2(w,\rho)\,h_{12}(w) \int_0^\infty L_1(v, w)\,h_{12}(v)$$

$$\times \int_v^\infty K_2(v,t)\,g_1(t)\,h_{13}(t)\,dt\,dv\,dw\,, \quad (6)$$

$$0 < \rho < a\,,$$

and

$$\int_0^\infty G(t,\rho)\,g_2(t)\,dt = \int_0^\infty g_2(t)\,h_{21}(\rho)\,h_{23}(t)$$

$$\times \int_\rho^\infty\!\!\int_t^\infty K_2(\rho,w)\,K_2(t,v)\,h_{22}(w)\,h_{22}(v)$$

$$\times L_2(v,w)\,dv\,dw\,dt$$

$$= h_{21}(\rho)\int_\rho^\infty K_2(\rho,w)\,h_{22}(w)\int_0^\infty L_2(v,w)\,h_{22}(v)$$

$$\times \int_0^v K_2(t,v)\,g_2(t)\,h_{23}(t)\,dt\,dv\,dw\,, \tag{7}$$

$$b < \rho < \infty\,,$$

where we have assumed that various orders of integration may be interchanged.

From equations (3) and (6) and the first part of (10.2.8), we derive the relation

$$h_{11}(\rho)\int_0^\rho K_2(w,\rho)\,[h_{12}(w)]^2 \int_w^\infty K_2(w,t)\,g_1(t)\,h_{13}(t)\,dt\,dw$$

$$= f_1(\rho) - h_{11}(\rho)\int_0^\rho K_2(w,\rho)\,h_{12}(w)\int_0^\infty L_1(v,w)\,h_{12}(v)$$

$$\times \int_v^\infty K_2(v,t)\,g_1(t)\,h_{13}(t)\,dt\,dv\,dw\,, \qquad 0 < \rho < a\,. \tag{8}$$

Similarly, from equations (4) and (7) and the second part of (10.2.8), we have

$$h_{21}(\rho)\int_\rho^\infty K_2(\rho,w)\,[h_{22}(w)]^2 \int_0^w K_2(t,w)\,g_2(t)\,h_{23}(t)\,dt\,dw$$

$$= f_2(\rho) - h_{21}(\rho)\int_\rho^\infty K_2(\rho,w)\,h_{22}(w)\int_0^\infty L_2(v,w)\,h_{22}(v)$$

equation (24) becomes

$$\frac{\partial^2 \psi}{\partial \rho^2} + \frac{1}{\rho}\frac{\partial \psi}{\partial \rho} + \frac{\partial^2 \psi}{\partial z^2} - \frac{4}{\rho^2}\psi = 0 . \tag{26}$$

Note that χ vanishes on the cylinder and is equal to $-\frac{1}{4}\tau\rho^4$ on S_1. Thus, we have the boundary value problem

$$\nabla^2(\psi \cos 2\varphi) = 0 \quad \text{in} \quad R , \tag{27}$$

$$\psi \cos 2\varphi = -\frac{1}{4}\tau\rho^2 \cos 2\varphi \quad \text{on} \quad S_1 , \tag{28}$$

$$\psi \cos 2\varphi = 0 \quad \text{on} \quad S_2 , \tag{29}$$

where R is the region between S_1 and S_2, while S_2 stands for the surface of the cylinder.

Following the method explained in Sections 6.4, 6.5, and 11.2, we can readily give an integral-equation formulation to this boundary value problem. The result is

$$-\frac{1}{4}\tau\rho^2 = \int_C [G_c^{(2)}(\rho,z;\rho_1,z_1) + 2\pi G_1^{(2)}(\rho,z;\rho_1,z_1)]\,\sigma(\rho_1,z_1)\,\rho_1\,ds, \tag{30}$$

where

$$\sigma(\rho_1,z_1) = \frac{1}{4\pi}\left(\frac{\partial \psi}{\partial n}\bigg|_- - \frac{\partial \psi}{\partial n}\bigg|_+\right),$$

and $\pi^{-1}G_0^{(2)}$ is the coefficient of $\cos 2(\varphi - \varphi_1)$ in the Fourier expansion of the free-space Green's function [cf. (6.4.30)] while $G_1^{(2)}$ is defined by equation (11.2.11) with $r = 2$. The curve C is the bounding curve of S_1 in the meridian plane, ds denotes the element of the arc length measured along C, and both (ρ,z) and (ρ_1,z_1) lie on C.

The next step is to introduce a small parameter ε in the problem and also to find the constant A occurring in the relation (11.1.10). Let a denote the maximum distance between two points of C and let $a \ll b$; then, $\varepsilon = a/b$. The constant A is found by substituting the expansion for the modified Bessel function I_2 in the expression for $G_1^{(2)}$. Then, it follows that

$$2\pi G_1^{(2)} = \rho^2 \rho_1{}^2 [A + G_3(\rho,z;\rho_1,z_1)] , \tag{31}$$

where

$$A = -\frac{3\pi}{64b^5}\beta_4, \qquad \beta^{2r} = \frac{2^{r+1}}{\pi(2r)!}\int_0^\infty v^{2r}\frac{K_2(v)}{I_2(v)}\,dv, \qquad (32)$$

and G_3 is of order $A\varepsilon^2$ (see Exercise 14).

Thereafter, the analysis is similar to the one given in Section 11.2. In fact, from (30), we have

$$-\tfrac{1}{4}\tau\rho^2 = \int_C G_0^{(2)}\sigma'\rho_1\,ds + \rho^2\int_C G_3\sigma'\rho_1{}^3\,ds, \qquad (33)$$

where

$$\sigma' = \sigma[1 + (4A/\tau)\int_C \rho_1{}^3\sigma\,ds]^{-1}. \qquad (34)$$

The relation (34), in turn, gives

$$\int_C \rho_1{}^3\sigma\,ds = \int_C \rho_1{}^3\sigma'[1 - (4A/\tau)\int_C \rho_1{}^3\sigma'\,ds]^{-1}\,ds. \qquad (35)$$

As in the case of the electrostatic problem of Section 11.2, $\sigma'\simeq\sigma_0$, where σ_0 arises in the integral equation

$$-\tfrac{1}{4}\tau\rho^2 = \int_C G_0^{(2)}\sigma_0\rho_1\,ds \qquad (36)$$

for the same crack problem in an infinite medium. Then, it follows from (35) that

$$\int_C \rho_1{}^3\sigma\,ds \simeq \int_C \rho_1{}^3\sigma_0[1 - (4A/\tau)\int_C \rho_1{}^3\sigma_0\,ds]^{-1}\,ds, \qquad (37)$$

that is, the solution for a crack in a cylinder may be found approximately from the corresponding solution in an infinite medium.

Finally, we evaluate E, the loss of potential energy due to a crack in an infinite cylinder, in terms of E_0, the loss of the corresponding quantity due to a crack in an infinite medium. The value of E is defined by the formula

$$E = (4\pi/\mu)\int_R \rho(\sigma_{z\varphi}^2 + \sigma_{\rho\varphi}^2)\,dV. \qquad (38)$$

Using the relations (23) and (25), and after a slight manipulation, the relation (38) becomes

$$E = -4\pi^2 \mu\tau \int_C \rho^3 \sigma \, ds$$

$$\simeq -4\pi^2 \mu\tau \int_C \rho^3 \sigma_0 [1 - (4A/\tau) \int_C \rho_1^{\,3} \sigma_0 \, ds]^{-1} \, ds \,, \qquad (39)$$

or

$$E \simeq E_0 [1 + (AE_0/\mu\tau^2 \pi^2)]^{-1} \,. \qquad (40)$$

11.5. THEORY OF DIFFRACTION

Finally, we use the method of this chapter for studying the theory of diffraction. Indeed, the method was first introduced for solving problems of this very theory. When we introduce a bounded obstacle B with surface S in a source-free region of the incident field u_i, then this field is disturbed. The total field is $u = u_i + u_s$, where u_s is the diffracted or scattered field, defined only in the exterior region R_e of S. Following the procedure of Section 6.6 and that of Example 1 of Section 6.7, we find that, for a perfectly soft body, we have the following boundary value problem:

$$\nabla^2 k^2 u_s + k^2 u_s = 0 \,, \qquad x \in R_e \,, \qquad (1)$$

$$u_s = -u_i \qquad \text{on} \quad S \,, \qquad (2)$$

and u_s satisfies the Sommerfeld radiation condition. The integral-equation representation formula for this problem, by the method and notation of Chapter 6, is

$$u_s(x) = -\int_S \sigma(\xi) E(x;\xi) \, dS \,, \qquad (3)$$

where $\sigma(x) = \partial u(x)/\partial n$ is the single-layer density. When x approaches a point on S, we have, in the limit,

$$u_i(x) = \int_S \sigma(\xi) E(x;\xi) \, dS \,, \qquad x, \xi \in S \,, \qquad (4)$$

which is a Fredholm integral equation of the first kind for evaluating the function $\sigma(\mathbf{x})$.

For a perfectly rigid obstacle, we have the Neumann boundary value problem consisting of the differential equation (1) and the boundary condition

$$\partial u_s/\partial n = -\partial u_i/\partial n \qquad \text{on} \quad S. \tag{5}$$

The integral representation formula that embodies (1) and (5) is

$$u_s(\mathbf{x}) = \int_S u(\xi)\,[\partial E(\mathbf{x};\xi)/\partial n]\,dS, \qquad \mathbf{x} \in R_e. \tag{6}$$

When $\mathbf{x}$ approaches a point on S, we use the relation (6.6.10) and obtain

$$u_s(\mathbf{x}) = \tfrac{1}{2}u(\mathbf{x}) + \int_S u(\xi)\,[\partial E(\mathbf{x};\xi)/\partial n]\,dS, \qquad \mathbf{x},\xi \in S \tag{7}$$

or

$$2u_i(\mathbf{x}) = \tau(\mathbf{x}) - 2\int_S \tau(\xi)\,[\partial E(\mathbf{x};\xi)/\partial n]\,dS, \qquad \mathbf{x},\xi \in S, \tag{8}$$

where $\tau(\mathbf{x}) = u(\mathbf{x})$ is the double-layer density.

Example.   Let us solve the problem of the diffraction of a plane wave by a soft sphere. It is convenient to take spherical polar coordinates (r, θ, φ). Thus, $\mathbf{x} = P = (r, \theta, \varphi)$, $\xi = Q = (a, \theta_1, \varphi_1)$, and $|\mathbf{x} - \xi| = R$.

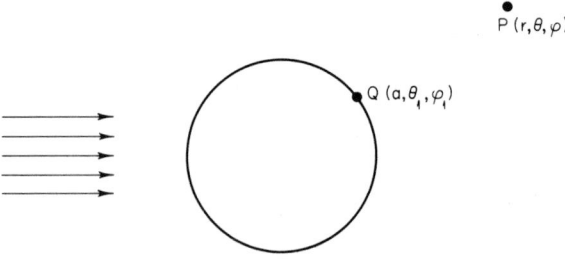

Figure 11.1

In this case (see Figure 11.1)

$$u_i(P) = e^{iak(\cos\theta)} = 1 + iak(\cos\theta) - \frac{a^2 k^2}{2}(\cos^2\theta)$$

$$- i\frac{a^3 k^3 \cos^3\theta}{6} + O(k^4). \tag{9}$$

$$E(P;Q) = \frac{e^{ikR}}{4\pi R} = \frac{1}{4\pi R} + \frac{ik}{4\pi} - \frac{k^2 R}{8\pi} - \frac{ik^3 R^2}{24\pi} + O(k^4). \tag{10}$$

$$\sigma(Q) = \sigma_0(Q) + k\sigma_1(Q) + k^2\sigma_2(Q) + k^3\sigma_3(Q) + O(k^4). \tag{11}$$

Substituting (9)–(11) in (4) and equating equal powers of k, we have

$$1 = \int_S \frac{\sigma_0(Q)}{4\pi R}\,dS, \tag{12}$$

$$ia\cos\theta = \int_S \frac{\sigma_1(Q)}{4\pi R}\,dS + \int_S \frac{i\sigma_0(Q)}{4\pi}\,dS, \tag{13}$$

$$-\frac{a^2}{2}\cos^2\theta = \int_S \frac{\sigma_2(Q)}{4\pi R}\,dS + \int_S \frac{i\sigma_1(Q)}{4\pi}\,dS - \int_S \frac{R\sigma_0(Q)}{8\pi}\,dS, \tag{14}$$

and so on. The integral equations (12)–(14) are of the same form as (11.3.79)–(11.3.82) and are easily solved by introducing Legendre polynomials. The solution is

$$\sigma_0 = 1/a, \qquad \sigma_1 = -i + 3iP_1(\cos\theta_1) = -i + 3i\cos\theta_1,$$
$$\sigma_2 = -\tfrac{1}{2}aP_0(\cos\theta_1) - \tfrac{5}{3}aP_2(\cos\theta_1) = \tfrac{1}{3}a - \tfrac{5}{2}a\cos^2\theta_1. \tag{15}$$

Consequently,

$$\sigma(Q) = \sigma(\theta_1)$$
$$= (1/a)[1 + i\varepsilon(3\cos\theta_1 - 1) + \varepsilon^2(\tfrac{1}{3} - \tfrac{5}{2}\cos^2\theta_1) + O(\varepsilon^3)], \tag{16}$$

where $\varepsilon = ak$ and we assume that $\varepsilon \ll 1$.

The quantities of physical interest in the theory of scattering are the far-field amplitude and the scattering cross section, which we now evaluate for the present example. Since the scattered wave is an outgoing wave and satisfies the radiation condition, we expect its far-field behavior to be given by

$$u_s(r, \theta, \varphi) \simeq A(\theta, \varphi) e^{ikr}/r \ . \tag{17}$$

Now, for large r,

$$R = |\mathbf{x} - \boldsymbol{\xi}| = (r^2 + a^2 - 2ar\cos\gamma)^{\frac{1}{2}}$$

$$= r[1 + (a^2/r^2) - (2a/r)\cos\gamma]^{\frac{1}{2}} \simeq (r - a\cos\gamma) \ , \tag{18}$$

where

$$\cos\gamma = \cos\theta\cos\theta_1 + \sin\theta\sin\theta_1\cos(\varphi - \varphi_1) \ .$$

Hence, from (3), we have

$$u_s(r, \theta, \varphi) \simeq A(\theta) e^{ikr}/r \ , \tag{19}$$

where

$$A(\theta) = -(a^2/4\pi) \int_0^\pi \int_0^{2\pi} e^{-i\varepsilon\cos\gamma}\sigma(\theta_1)\sin\theta_1 \, d\theta_1 \, d\varphi_1 \ , \tag{20}$$

is the required far-field amplitude. The next step is to use the expansions of the terms inside the integral sign of (20), integrate, and obtain

$$A(\theta) = -a[1 - i\varepsilon + \varepsilon^2(\cos\theta - \tfrac{2}{3}) + O(\varepsilon^3)] \ . \tag{21}$$

The value of the scattering cross section is given by the formula

$$\text{S.C.S.} = 2\pi \int_0^\pi |A(\theta)|^2\sin\theta \, d\theta = 4\pi a^2[1 - (\varepsilon^2/3) + O(\varepsilon^4)] \ . \tag{22}$$

EXERCISES

1. Show that formula (11.2.19) for the capacity is correct to order ε^6.

2. Instead of spherical coordinates (11.2.12), take oblate-spheroidal coordinates

$$z = ae\xi \ , \qquad \rho = ae[(1 - \xi^2)(1 + \zeta^2)]^{\frac{1}{2}} \ ,$$

and solve the electrostatic potential problem for the case of an oblate spheroid placed symmetrically inside a grounded infinite cylinder of radius b. Show that the capacity C of this condenser is

$$C = \frac{ae}{\sin^{-1}e}\left\{1 - \frac{2\varepsilon e}{\pi \sin^{-1}e}\right.$$

$$\times \left[I(0) + \frac{1}{9}(e\varepsilon)^2 \frac{I(2)}{9} + \frac{2}{225}(e\varepsilon)^4 I(4)\right]^{-1}\right\} + O(\varepsilon^6),$$

where the quantities $I(2m)$ are defined by (11.2.20).

Hint: Use the expansion

$$I_0(p\rho)\,e^{ipz} = \sum B_n(p)\,P_n(\xi)\,P_n(i\zeta)\,,$$

where the P_n are Legendre polynomials.

3. Following a procedure similar to that in Exercise 2, solve the electrostatic potential problem for a prolate spheroid.

4. Solve Exercises 2 and 3 when the spheroids are placed between two grounded parallel plates.

5. Derive formula (11.3.43).

6. Substitute the values of the densities σ_0, σ_1, σ_2, and σ_3 as given by (11.3.65) in (11.3.51) and obtain the velocity field.

7. By using formula (11.3.61), obtain the frictional torque experienced by a sphere of radius a which is oscillating in a viscous fluid.

8. Do the same as in Exercise 7 for a spheroid and a disk.

9. Calculate the $O(\mathscr{R})$ term in formula (11.3.72) explicitly and use this expression to deduce an approximate formula for the drag experienced by a solid in Oseen flow. Illustrate this formula for the case of a sphere.

10. Use the values of the charge densities as given by the formulas (11.3.86) and evaluate the velocity field set up in Oseen flow when a sphere is rotating uniformly and is placed symmetrically inside a circular cylinder.

11. Find the torque experienced by a sphere which is rotating uniformly in Oseen flow and is bounded by a pair of parallel walls $z = \pm c$. Evaluate also the velocity field.

12. Extend the analysis of the steady Oseen flow in the text to the case of the steady-state vibrations of axially symmetric solids in Oseen flow.

13. Use the analysis of Section 11.4 and discuss the problem of a spherical inclusion in an elastic half-space.

14. Derive the integral equation (11.4.30). Prove that the quantity G_3 occurring in the relation (11.4.31) is of order $A\varepsilon^2$. Also prove that σ' occurring in the relation (11.4.33) is such that $\sigma' = \sigma_0[1 + O(\varepsilon^7)]$.

15. Start with the integral equation (11.5.8) and solve the problem of diffraction of a plane wave by a perfectly rigid sphere.

16. Solve the problem of diffraction of a plane wave by a perfectly soft and by a rigid circular disk. Also solve the dual problem of diffraction by an aperture.

17. Use the approximation

$$\frac{i}{4} H_0^{(1)}(kR) = -\frac{1}{2\pi}(q + \log R) + \frac{k^2 R^2}{8\pi}(q - 1 + \log R) + \log\tfrac{1}{2}k\,\{O(k^4)\},$$

where $q = \gamma + \log\tfrac{1}{2}k - \tfrac{1}{2}\pi i$ (γ is Euler's constant), and solve the problem of diffraction of a plane wave by a soft and by a rigid circular cylinder.

APPENDIX

A.1. To prove the identity

$$\int_a^s \int_a^{s_n} \cdots \int_a^{s_3} \int_a^{s_2} F(s_1) \, ds_1 \, ds_2 \cdots ds_{n-1} \, ds_n$$

$$= [1/(n-1)!] \int_a^s (s-t)^{n-1} \, F(t) \, dt \,, \tag{1}$$

we begin with the formula

$$\frac{d}{ds} \int_{A(s)}^{B(s)} f(s,t) \, dt = \int_A^B \frac{\partial f(s,t)}{\partial s} \, dt + f[s, B(s)]\frac{dB}{ds} - f[s, A(s)]\frac{dA}{ds} \,, \tag{2}$$

for differentiation of an integral involving a parameter. Let us apply this formula for the differentiation of the function $I_n(s)$ defined by the relation

$$I_n(s) = \int_a^s (s-t)^{n-1} F(t) \, dt \,, \tag{3}$$

where n is a positive integer and a is a constant. For this purpose, set

$$f(s, t) = (s - t)^{n-1} F(t)$$

in (2). The result is

$$dI_n/ds = (n-1) \int_a^s (s-t)^{n-2} F(t)\, dt + [(s-t)^{n-1} F(t)]_{t=s}$$

$$= (n-1) I_{n-1}, \qquad n > 1. \tag{4}$$

For $n = 1$, we have directly from (3)

$$dI_1/ds = F(s). \tag{5}$$

From the recurrence relation (4), we obtain

$$d^k I_n/ds^k = (n-1)(n-2) \cdots (n-k) I_{n-k}, \qquad n > k, \tag{6}$$

which for $k = n-1$ becomes

$$d^{n-1} I_n/ds^{n-1} = (n-1)!\, I_1(s). \tag{7}$$

Differentiating (7) and using (5) results in the equation

$$d^n I_n/ds^n = (n-1)!\, F(s). \tag{8}$$

Furthermore, from the relations (3), (6), and (7) it follows that $I_n(s)$ and its first $n-1$ derivatives all vanish when $s = a$. Hence, equations (5) and (8) yield

$$I_1(s) = \int_a^s F(s_1)\, ds_1,$$

$$I_2(s) = \int_a^s I_1(s_2)\, ds_2 = \int_a^s \int_a^{s_2} F(s_1)\, ds_1\, ds_2, \tag{9}$$

$$I_n(s) = (n-1)! \int_a^s \int_a^{s_n} \cdots \int_a^{s_3} \int_a^{s_2} F(s_1)\, ds_1\, ds_2 \cdots ds_{n-1}\, ds_n. \tag{10}$$

Combining (3) and (10), we have the required identity (1).

A.2. It can be easily proved by the complex integration method that

$$\int_0^\infty p\left(\frac{p}{\gamma}-1\right)J_\mu(pv)J_\mu(p\rho)\,dp = \begin{cases} i\int_0^k \frac{p^2}{(k^2-p^2)^{1/2}}\,H_\mu^{(1)}(pv)J_\mu(p\rho)\,dp\,, \\[2mm] \qquad v\geq\rho\,, \\[4mm] i\int_0^k \frac{p^2}{(k^2-p^2)^{1/2}}\,J_\mu(pv)\,H_\mu^{(1)}(p\rho)\,dp\,, \\[2mm] \qquad \rho\geq v\,, \end{cases} \tag{11}$$

where

$$\gamma = \begin{cases} -i(k^2-p^2)^{1/2}\,, & k\geq p\,, \\ (p^2-k^2)^{1/2}\,, & p\geq k\,, \end{cases} \tag{12}$$

and $v,\rho > 0$.

Suppose that $v\geq\rho$ and let the complex plane be $s = \sigma + i\tau$. Integrating

$$\{[s^2/(s^2-k^2)^{1/2}] - s\}\,H_\mu^{(1)}(sv)J_\mu(s\rho)$$

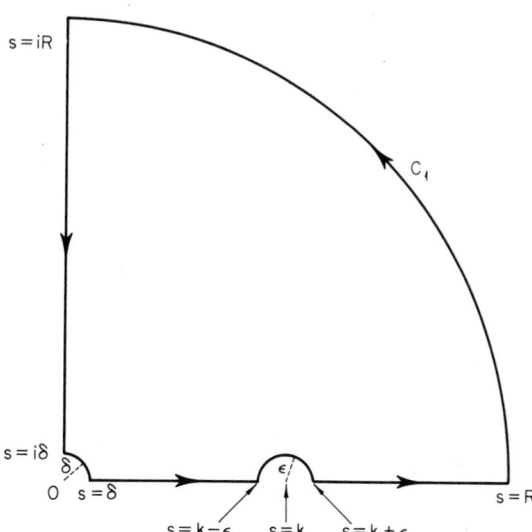

Figure A.1

around a contour C_1 in the upper right-hand quadrant passing over the branch point $s = k$, as shown in Figure A.1, we get

$$\oint_{C_1} \{[s^2/(s^2 - k^2)^{1/2}] - s\}\, H_\mu^{(1)}(sv)\, J_\mu(s\rho)\, ds = 0 ,$$

because there are no singularities within this contour. If we let $\delta, \varepsilon \to 0$ and $R \to \infty$, the contributions from the corresponding arcs tend to zero. Hence,

$$\int_0^k \left(\frac{\sigma^2}{i(k^2 - \sigma^2)^{1/2}} - \sigma \right) H_\mu^{(1)}(\sigma v)\, J_\mu(\sigma\rho)\, d\sigma + \int_k^\infty \left(\frac{\sigma^2}{(\sigma^2 - k^2)^{1/2}} - \sigma \right)$$

$$\times H_\mu^{(1)}(\sigma v)\, J_\mu(\sigma\rho)\, d\sigma$$

$$+ i \int_\infty^0 \left(\frac{-\tau^2}{i(\tau^2 + k^2)^{1/2}} - i\tau \right) H_\mu^{(1)}(i\tau v)\, J_\mu(i\tau\rho)\, d\tau = 0 . \tag{13}$$

Similarly, integrating

$$\{[s^2/(s^2 - k^2)^{1/2}] - s\}\, H_\mu^{(2)}(sv)\, J_\mu(s\rho)$$

around a contour C_2 in the lower right-hand quadrant and passing under the branch point $s = k$ gives

$$\int_0^k \left(\frac{\sigma^2}{-i(k^2 - \sigma^2)^{1/2}} - \sigma \right) H_\mu^{(2)}(\sigma v)\, J_\mu(\sigma\rho)\, d\sigma + \int_k^\infty \left(\frac{\sigma^2}{(\sigma^2 - k^2)^{1/2}} - \sigma \right)$$

$$\times H_\mu^{(2)}(\sigma v)\, J_\mu(\sigma\rho)\, d\sigma$$

$$- i \int_\infty^0 \left(\frac{\tau^2}{i(\tau^2 + \sigma^2)} + i\tau \right) H_\mu^{(2)}(-i\tau v)\, J_\mu(-i\tau\rho)\, d\tau = 0 . \tag{14}$$

Now using the relation $H_\mu^{(1)}(i\tau v)\, J_\mu(i\tau\rho) = -H_\mu^{(2)}(-i\tau v)\, J_\mu(-i\tau\rho)$ and adding equations (13) and (14), we obtain, for $v \geqslant \rho$,

$$\int_k^\infty \left(\frac{\sigma^2}{(\sigma^2-k^2)^{\frac{1}{2}}} - \sigma \right) J_\mu(\sigma v) J_\mu(\sigma\rho)\, d\sigma$$

$$= \int_0^k \sigma J_\mu(\sigma v) J_\mu(\sigma\rho)\, d\sigma - \int_0^k \frac{\sigma^2}{(k^2-\sigma^2)^{\frac{1}{2}}} Y_\mu(\sigma v) J_\mu(\sigma\rho)\, d\sigma, \quad (15)$$

where $Y\mu$ is the Bessel function of the second kind. It follows from (13) and (15) that the left-hand side of (11) is

$$\int_0^\infty p\left(\frac{p}{\gamma}-1\right) J_\mu(pv) J_\mu(p\rho)\, dp$$

$$= \int_0^k \left(\frac{ip^2}{(k^2-p^2)^{\frac{1}{2}}} - p \right) J_\mu(pv) J_\mu(p\rho)\, dp$$

$$+ \int_k^\infty \left(\frac{p^2}{(p^2-k^2)^{\frac{1}{2}}} - p \right) J_\mu(pv) J_\mu(p\rho)\, dp$$

$$= i \int_0^k \frac{p^2}{(k^2-p^2)^{\frac{1}{2}}} [J_\mu(pv) + iY_\mu(pv)] J_\mu(p\rho)\, dp$$

$$= i \int_0^k \frac{p^2}{(p^2-k^2)^{\frac{1}{2}}} H_\mu^{(1)}(pv) J_\mu(p\rho)\, dp, \qquad v \geqslant \rho,$$

which proves the first part of formula (11). The second part follows in a similar fashion.

BIBLIOGRAPHY

1. Bocher, M., "An Introduction to the Study of Integral Equations" (Cambridge Tracts in Mathematics and Mathematical Physics, No. 10). Cambridge Univ. Press, 1909.
2. Bückner, H., "Die practische Behandlung von Integralgleichungen" (Ergebnisse der angewandten Mathematik, No. 1). Springer, New York, 1952.
3. Carrier, G. F., Krook, M., and Pearson, C. F., "Functions of a Complex Variable." McGraw-Hill, New York, 1966.
4. Courant, R., and Hilbert, D., "Methods of Mathematical Physics," Vols. I and II. Wiley, New York, 1953, 1962.
5. Haberman, W. L., and Harley, E. E., Numerical evaluation of integrals containing modified Bessel functions, David Taylor Model Basin Report No. 1580, 1964.
6. Hilderbrand, F. B., "Methods of Applied Mathematics," 2nd ed. Prentice-Hall, Englewood, New Jersey, 1965.
7. Irving, J., and Mullineux, N., "Methods in Physics and Engineering." Academic Press, New York, 1959.
8. Jain, D. L., and Kanwal, R. P., "Mixed Boundary Value Problems of Mathematical Physics." In preparation.
9. Kellog, O. D., "Foundations of Potential Theory." Dover, New York, 1953.
10. Lovitt, W. V., "Linear Integral Equations." Dover, New York, 1950.
11. Mikhlin, S. G., "Integral Equations." Pergamon Press, Oxford, 1957.
12. Morse, P. M., and Feshback, H., "Methods of Theoretical Physics," Vols. I and II. McGraw-Hill, New York, 1953.

13. Mushkhelishvili, N. I., "Singular Integral Equations," 2nd ed. P. Noordhoff N.V., Gronignen, Holland, 1946.
14. Peterovskii, I. G., "Integral Equations," Grylock Press, New York, 1957.
15. Pogorzelski, W., "Integral Equations and Their Applications," Vol. 1. Pergamon Press, Oxford, 1966.
16. Smirnov, V. I., "Integral Equations and Partial Differential Equations." Addison-Wesley, Reading, Massachusetts, 1964.
17. Sneddon, I. N., "Fourier Transforms." McGraw-Hill, New York, 1951.
18. Sneddon, I. N., "Mixed Boundary Value Problems in Potential Theory." Wiley, New York, 1966.
19. Stakgold, I., "Boundary Value Problems of Mathematical Physics," Vols. I and II. Macmillan, New York, 1967, 1968.
20. Tricomi, F. G., "Integral Equations." Wiley, New York, 1957.
21. Watson, G. N., "A Treatise on the Theory of Bessel Functions." Cambridge Univ. Press, London and New York, 1962.
22. Weinberger, H. F., "A First Course in Partial Differential Equations." Blaisdell, New York, 1965.
23. Yosida, K., "Lectures on Differential and Integral Equations." Wiley, New York, 1960.

INDEX